Wild About Johannesburg

ALL-IN-ONE GUIDE TO
COMMON ANIMALS & PLANTS
OF GARDENS, PARKS AND NATURE RESERVES

Duncan Butchart

SOUTHERN
BOOK PUBLISHERS

*D*edicated to my parents, who
encouraged my interest in the
natural world.

Acknowledgements

Many friends and colleagues have assisted me in the compilation of this book and their
support is greatly appreciated. Valuable comments on the introductory sections were
provided by Ann Cameron, Jane Carruthers and Lex Hes. Individual sections were read
by specialists in their fields: John and Sandie Burrows (plants), Buster Culverwell
(reptiles), Vincent Carruthers (frogs) and Dr Paul Skelton (fishes); any errors are,
however, of my own making. The names of the photographers who supplied material
additional to my own are listed alongside their respective pictures. I am particularly
grateful to Lex Hes, Vincent Carruthers, Colin Bell, Paul Skelton, Beth Peterson and
Chris and Tilde Stuart for their generous provision of photographs, and to James
Marshall for the loan of photographic equipment. Brendan Ryan and John Carlyon kindly
permitted the use of photographs at a reduced rate. Logistical support and assistance
during research and photographic excursions to Johannesburg was provided by my
parents-in-law Cecil and Shelagh Peterson, and by Beth Peterson and Ian Sutherland.
My exploration of the animals and plants of the Johannesburg area began in 1970 in the
hills surrounding my parents' home in Mondeor. In those early years, I was fortunate to
receive guidance and encouragement from John Freer and John Ledger. I am grateful to
Southern Book Publishers for their belief in the 'Wild About' series, and in particular to
Louise Grantham for her enthusiasm and Mike Thayer for his assistance. Finally, I thank
my wife Tracey for her comments on various drafts of the text, and for her love, support
and encouragement.

Cover photographs: Greater Flamingo (Lex Hes); Rock Elephant Shrew (C & T Stuart); Rinkhals
(WD Haacke); Large Witchweed, Sweet Thorn (Duncan Butchart); Red Toad (Vincent Carruthers).
Photograph opposite: Lorna Stanton (ABPL).

ISBN 1 86812 595 5

First edition, first impression 1995

Published by
Southern Book Publishers (Pty) Ltd
PO Box 3103, Halfway House 1685, South Africa

Design and typesetting by Groundhog Graphics, Nelspruit
Reproduction by Hirt & Carter, Cape Town
Printed and bound by National Book Printers, Drukkery Street, Goodwood, Western Cape

Contents

INTRODUCTION

This book is intended as a simple introduction to the birds, mammals, plants and other forms of wildlife found in the greater Johannesburg area, and as a compact field guide to the identification of the more common and conspicuous species. It is aimed as much at the casual garden naturalist as it is at the enthusiast who wishes to explore the whole area. The guide does not attempt to be comprehensive, as there already exists a wide variety of excellent books which focus in detail on individual groups of southern African animals and plants. For the same reason, this is not a 'where to go' book, although a map showing places of interest and a list of useful addresses are included.

In compiling this work, my goal has been twofold. Firstly, to produce a guide which stimulates an interest in the surprising variety of wildlife coexisting with mankind in the vicinity of South Africa's largest city. Secondly, to deal with the full spectrum of wildlife – from mammals and birds to insects and plants – and to emphasise the importance of specific habitats so that the user may gain an understanding and appreciation of the interdependence of all living things. This understanding is critical to the effective conservation of all life forms – including mankind – on our fragile planet.

The area covered by this guide is featured on the map on pp. 6 and 7; between Randfontein, Fourways, Springs and the Klip River – an area encompassing some 4 800 km^2. Although not featured on the map, this guide may also be useful in the Suikerbosrand Nature Reserve near Heidelberg. The flora and fauna of the Magaliesberg is typical of the bushveld/woodland biome, and is not covered here.

Few parts of South Africa have had their environment so radically altered as Johannesburg and its surroundings. Prior to the arrival of gold miners and the consequent settlement of increasing numbers of people, the landscape was dominated by open grassland, punctuated by rocky outcrops and ridges, and was largely devoid of trees. Today, the ridges are still there but the expansion of suburbs has transformed much of the area into imitation woodland and even forest. This man-made network of gardens, avenues and parks is now home to species such as the Olive Thrush and the notorious 'Parktown Prawn' (King Cricket) – both typical of damp forests. The abundance of introduced plants with nectar-rich flowers and juicy fruits has proved a drawcard for previously unrecorded birds such as the Grey Lourie and Rameron Pigeon.

But as trees have been planted, and open spaces have become wooded, so the original grassland inhabitants have retreated. A hundred or more years ago, Springbok and Blue Crane would have been at home in the areas now known as Rosebank and Randburg. These larger mammals and bigger birds were, of course, the first to disappear as their habitat needs conflicted with human development.

Johannesburg is one of the few large cities in the world which is not situated alongside a major waterway of some kind. But despite the absence of any large river, important and fascinating wetlands do exist, most notably on the East Rand. In recent years, some moves have been made to clean up the streams which drain north and south off the Witwatersrand, but this is an ongoing task, which requires the full support of local communities and industry. Nevertheless, the creation of numerous artificial water bodies has made the area very attractive to waterfowl and other forms of water-dependent life. The sight of large flocks of Sacred Ibis – flying in V-formation to their evening roost at Zoo Lake – is an unexpected but impressive wildlife spectacle in the heart of the city.

The largest natural area is the Klipriviersberg Nature Reserve, south of the city, which incorporates unspoiled grassy hills, thicketed ravines and thorn tree woodland. Other sizeable natural areas are the Marievale Bird Sanctuary, Rhenosterpoort Private Nature Reserve, Diepsloot Nature Area and the Krugersdorp Game Reserve. Rondebult Bird Sanctuary is renowned as a waterfowl refuge, while the magnificent Witwatersrand National Botanical Garden contains a sizeable area of natural vegetation in addition to cultivated flora.

Residents and visitors to Johannesburg will probably be surprised by the diversity and tenacity of the wildlife. Given a chance, many species can survive alongside people, and there is great scope for intensified urban conservation. A combined effort by town planners, gardeners, industrialists, teachers and school children can ensure that wild species are encouraged and accommodated in and around the city.

HOW TO USE THIS GUIDE

The animals and plants

In contrast to more comprehensive field guides, where the user aims to identify a species once it has been encountered, this book can also be used in the reverse manner. Because only the more common species are featured, it is possible for the user to **actively seek** them out. This approach is particularly useful with trees and other plants, which pose identification problems for many people.

For comparative purposes, species which share similar features are arranged together, even though this sequence may differ from standard reference works. In addition to the animals and plants featured, mention is also made of other, less common species with which they might be confused, these similar species (ss) are listed in small type. Although bats, rodents, frogs and others are not commonly seen – or noticed – those featured may be fairly easily **found** if one knows where to look. For invertebrates other than butterflies, no attempt has been made to single out particular species, but rather to facilitate the identification of the group or family to which they belong.

Alien (non-indigenous) species are included only if they have become naturalised and have self-propagating 'wild' (feral) populations; they are marked with an asterisk (*). Many of these aliens – particularly the plants – have become extremely damaging to their indigenous counterparts, and to the environment as a whole, as they are able to out-compete the local flora. This is due largely to the absence of their natural parasites and predators (which were not introduced alongside them). The identification of these alien plants and animals is the first step towards their eradication.

The names of species follow those of the most recent authoritative publications, but in line with ornithological publications – and in order to standardise terminology in this guide – hyphens have been eliminated from all double-barrelled common names. The scientific names of trees and other plants are used ahead of common names, as the latter often differ from region to region, and frequently relate species to families to which they do not belong. The name of the family to which each featured plant belongs is included for this reason. The Afrikaans names of indigenous plants have been included where possible as these are often well-known and descriptive. Scientific terms have been kept to an absolute minimum but could not be avoided altogether; an abbreviated glossary of these terms is provided on p. 123.

At the beginning of each section introductory notes provide general hints on identification, and the recognised reference books and more detailed field guides are listed. In addition to this, a list of books for further reading is provided on p. 122.

The habitats

Plants and animals have evolved under specific circumstances, in particular habitats. Becoming familiar with the habitat requirements and preferences of various species is often a fundamental aspect of nature study. Although the Johannesburg landscape is very fragmented, seven distinctive habitats, from man-made suburban gardens to wetlands, are identified and discussed. Recognition of the habitat that you are in is an important aspect of identification as a large number of species are specific to certain vegetation or soil types.

The illustrations

The photographs which accompany each species have been chosen for their ability to best demonstrate the key identification features. Where male and female of one species differ in appearance only the male is illustrated and the female described in the accompanying text. In most cases plants have been photographed in close-up, in their most distinctive, and therefore obvious, phase. Some difficult-to-photograph species have been illustrated with colour paintings.

WITPOORTJIE FALLS, WITWATERSRAND NATIONAL BOTANICAL GARDEN

MAP OF REGION
and key to places of interest

1. Witwatersrand National Botanical Garden
Malcolm Street, Poortview, Roodepoort. Open daily. Tel. (011) 958 1750
2. Melville Koppies Nature Reserve
Judith Road, opposite Marks Park Sports Club. Open only on the 3rd
Sunday each month. Tel. (011) 646 2000
3. Emmarentia Dam/JHB Botanical Garden
Louw Geldenhuis and Olifants Roads, Emmarentia. Open daily.
Tel. (011) 782 7064
4. Delta Park/ Florence Bloom Bird Sanctuary
1st Avenue, Victory Park. Open daily. Tel. (011) 407 6801
5. Braamfontein Spruit Trail
Informal ramble along the riverside park system from Parktown to
Rivonia. Tel. (011) 803 9132
6. Kelland Bird Sanctuary
Pitsani Road, Kelland, Randburg. Open only on Sundays.
Tel. (011) 884 0180
7. Melrose Bird Sanctuary
Melrose Avenue, Birdhaven. Tel. (011) 407 6801
8. Zoo Lake/Johannesburg Zoo
Jan Smuts Avenue, Parkview. Open daily. Tel. (011) 646 2000

9. The Wilds
Houghton Drive, Houghton Estate. Open daily. Tel. (011) 407 6801
10. Gilooly's Farm
Boeing Avenue, Bedfordview. Open daily. Tel. (011) 407 6111
11. Rietfontein Ridge
Sandton. Access only by prior arrangement. Tel. (011) 803 9300/1
12. Lone Hill Koppies
Lone Hill Estate. Open 2nd Saturday each month. Tel. (011) 704 263
or 803 9300/1
13. Norscot Koppies
Fourways. Open 2nd Saturday and Sunday each month.
Tel. (011) 704 2632 or 803 9300/1
14. Cumberland Bird Sanctuary
Cumberland Road, Bryanston. Open weekends and public holidays.
Tel. (011) 706 1936 or 803 9300
15. Diepsloot Nature Area
R511, north of Fourways. Access only by prior arrangement.
Tel. (011) 728 7373
16. Kloofendal Nature Reserve
Galena Avenue, Roodepoort. Open daily. Tel. (011) 472 1439

HALFWAY HOUSE

NORTHERN PROVINCE

NORTH WEST

PRETORIA

GAUTENG

JOHANNESBURG

MPUMALANGA

Vaal River

FREE STATE

KEMPTON PARK

EDENVALE

RAND

10

28

26

BENONI

BOKSBURG

27

29

BEDFORDVIEW

BRAKPAN

GERMISTON

25

SPRINGS

KATLEHONG

Natalspruit

24

NIGEL

N
W E
S

Florida Lake
berg Road, Roodepoort. Tel. (011) 472 1439
Helderkruin Trail/ Little Falls
mal 3 km trail. Crous Drive, Roodepoort. Tel. (011) 472 1439
Sterkfontein Caves
tain Road, 10 km north of Krugersdorp. Open from Thursday to
ay. Tel. (011) 956 6342
Krugersdorp Game Reserve
west of Krugersdorp. Open daily. Tel. (011) 660 1076
Con Joubert Bird Sanctuary
gbek Road, Randfontein. Tel. (011) 486 0938 or 693 3826
Donaldson Dam
off R29, Westonaria Mine. Tel. (011) 753 0266
Klipriviersberg Nature Reserve
ss via Columbine Avenue into Bellefield (western end) or
arvan (eastern end), Mondeor. Open daily. Tel. (011) 646 2000
Marievale Bird Sanctuary
ne R42 north of Nigel. Open daily. Tel. (011) 739 2411
Rondebult Bird Sanctuary
iston. Off the R554 east of Alrode. Tel. (011) 871 7355

26. Rynfield Dam/C.R. Swart Park
Sarel Cilliers Avenue, Rynfield, Benoni. Open daily. Tel. (011) 845 1650
27. Korsman Bird Sanctuary
Kilfenora Road, Benoni. Access restricted. Tel. (011) 845 1650
28. Carlos Rolfe's Bird Sanctuary
Kelly Road, Jet Park. Access restricted. Tel. (011) 486 0938
29. Leeupan Bird Sanctuary
Tom Jones Street (R23), Benoni. Open daily. Tel. (011) 845 1650
30. Kareebosrand Conservancy
Hennops Hills. Access by prior arrangement. Tel. (011) 886 0123
31. Rhenosterpoort Private Nature Reserve
Broederstroom. Access by prior arrangement. Tel. (01215) 51164
32. Rhino Nature Reserve
Kromdraai. Open daily. Tel. (011) 957 0039
33. Wonder Cave
Kromdraai. Open daily. Tel. (011) 957 0240
34. Kromdraai Conservancy
Kromdraai. Access by prior arrangement. Tel. (011) 957 0008

GEOLOGY AND TOPOGRAPHY

The lie-of-the-land, or general topography, is a direct result of the geological history of an area. Geological processes are often difficult to grasp by the non-specialist, however, and few people show much interest in the rocks under their feet. This is unfortunate, as an examination of rock type and structure reveals the processes responsible for the formation of the land around us.

It is way beyond the scope of this book to attempt to introduce the principles of geology, but one remarkable fact will come as such a surprise to most that it alone is sure to trigger an increased interest in the topography of the area: present-day Johannesburg was once under the sea!

A general picture of the geology of the region can perhaps best be gained by first imagining a layered cake with a thick, firm base, this being a mass of granitic rock laid down some 3 200 million years ago when the continent was forming. Interspersed with this are intrusions of so-called greenstones – lavas resulting from volcanic activity. On top of this base is a more recent accumulation of sedimentary rocks, deposited between 2 800 and 2 400 million years ago at a time when the ocean reached into the area. These sedimentary rocks are comprised mostly of shale and quartz, and include reefs rich in gold.

The great weight of this sedimentary layer caused the land mass to sag and tilt, forcing the ocean to retreat, and resulted in the Witwatersrand Basin being raised some 1 800 metres above sea level. A wonderful example of the raised and tilted 'layered cake' can be seen alongside the Witpoortjie Falls in the Witwatersrand Botanical Garden. Here the layers of pale quartz and dark shale are prominent.

The more recent, and slowly on-going, process of erosion – caused by the weathering of rocks by water, wind and ice – has created the gently undulating landscape of today. The most dramatic examples of this erosion are to be seen at Lone Hill and Norscot Koppies, where the ancient granitic rocks have been exposed. Elsewhere, movements within the crust of the Earth may have been responsible for raised land forms such as Randpark Ridge.

Interestingly, the Witwatersrand ridge acts as a watershed between two oceans. Rain falling in Parktown drains into the Braamfontein Spruit, Jukskei, Crocodile, then Limpopo rivers to spill out into the Indian Ocean in Mozambique, while rain falling in Mondeor drains into the Bloubos Spruit, Klip, Vaal, then Orange rivers to enter the Atlantic Ocean in the Northern Cape.

CLIMATE

Situated towards the northern end of South Africa's vast Highveld plateau, and at an average altitude of 1 500 metres above sea level, the Johannesburg area has an enviable climate – warm to hot summers with a fairly high rainfall, and cool to cold winters with little or no rain and dazzling blue skies. Midwinter temperatures often drop below freezing, however, and the regular frost may be severe. Snow falls are a rare, but not unknown, phenomenon.

This general pattern of temperature shows local variation depending on the topography and aspect of a given area. Valleys and wetlands are much cooler at night and more prone to frost than higher-lying places, while north-facing hill slopes are warmer, dry out more quickly after rain, and support more drought-tolerant plants. Southern hill slopes, angled away from the sun's rays and in shadow for a greater part of each day, retain moisture for longer and allow for the growth of denser vegetation and taller trees.

Within the suburbs, micro habitats, which insulate against the heat of the day and cool of the night, are created through the close planting of trees. Icy winds often blow across the Highveld during the winter, with August being a particularly windy month, but well-wooded suburbs are often spared their full force.

Much of the rain comes in the form of violent thundershowers, and Johannesburg has one of the world's highest incidences of lightning strikes. The average annual rainfall ranges between 700 and 800 mm. With the extensive surface areas of concrete and tar now present, run-off into the north-flowing streams may reach flood proportions after heavy downpours. With more water entering rivers such as the Jukskei than ever before, banks are being increasingly eroded and large riverine trees swept away. Cycles of prolonged drought – lasting for three to four years – are a natural phenomenon.

IDENTIFYING AND WATCHING WILDLIFE

Perhaps the best advice that can be given to any aspiring naturalist, birdwatcher or plant enthusiast, is to **first get to know the commoner species which live in your own garden or nearby park**. Since this book is aimed at introducing and identifying the species which fall into this category, it is hoped that its use will create a solid foundation for more advanced observation and study.

Becoming familiar with the birds which visit your garden will ensure that when you **do** venture further afield the more unusual species will stand out more readily from those with which you have become acquainted. This approach applies to all other groups of wildlife, although for plants you will need to start your studies by repeatedly visiting a patch of undisturbed land or a botanical garden.

Consideration for wildlife should always take priority. Move slowly and quietly so as not to disturb the animals you are watching. Getting close to a wild animal is a hollow achievement if you disrupt its life in the process.

Although casual observation of wildlife, and list-making of species recorded, is a pleasurable and relaxing pursuit, the more detailed observation of nature can be a perpetual journey of discovery and enjoyment. The three prerequisites for becoming a good field naturalist are preparation, patience and dedication.

Preparation

To begin with, you should make yourself familiar with the species which you are likely to see in a particular area by studying books in advance. In this way, you can anticipate finding certain plants and animals, and eliminate others on the basis of their known distribution. A wide variety of books, magazines, tapes, evening courses presented by recognised experts and video identification aids are available to help and inform aspirant and experienced naturalists.

With some thoughtful planning, your own garden can be turned into a haven for wildlife. The growing of plants indigenous to your area is the first step towards recreating natural food-chains. Artificial food provision at bird-tables may encourage a host of birds and a birdbath will be a big attraction, especially in winter.

When venturing out to look for wildlife, a notebook, sun hat, walking shoes and neutral-coloured clothes are all recommended. A good pair of binoculars is essential, while a sketch-book, camera or portable tape recorder will enable you to document interesting observations.

Regrettably, it has to be said that walking about in the Johannesburg area (as in many other large cities) can pose safety problems. Always travel in pairs or small groups, and let a friend or family member know where you are going and when you expect to be back.

Patience

Simply sitting in a well-chosen place is one of the best ways to view wildlife. Providing that you remain still and quiet, birds and other animals will often accept you as a non-threatening part of their environment and may carry on their activities while you watch. Consider, for example, how the birds in your own garden are visible and active at times when all is quiet, but seem to vanish when cars come and go, or when visitors arrive. Ponds and streams are great places to sit in wait for wildlife, as there is almost always something of interest happening, whether it be the courtship dance of dragonflies or the sudden arrival of a kingfisher. 'Sit still, look long, and hold yourself quiet', reads the inscription on a certain wildlife park bench. Sound advice indeed.

Dedication

To **really** get to know the wildlife in your surroundings you will need to make nature study more than just a casual pastime. Getting out at all times of the day, and in all weather, can be uncomfortable but may be very rewarding. Although it is certainly true that birds are most active in the early hours, interesting observations may be made at unexpected times. Being out in the rain and wind may not produce long lists of species seen, but you could be rewarded with views of a Rock Elephant Shrew dashing from its burrow to snap up emerging termites, or of a Greater Kestrel hovering in a raging wind above your head. Such out-door activity will also give you a better understanding of the environ-mental conditions which shape the behaviour of animals. Venture into your own garden after dark with a torch, and search flowering plants for nocturnal spiders and insects, or damp corners for frogs.

Instead of just making a list of species seen, keep a diary of your interesting observations. This is an invaluable way of increasing your knowledge and gathering information for later comparison. It is worth noting that many of nature's secrets are unravelled by amateur naturalists rather than by professional biologists. Consider your field guides as working tools, not sacred tomes, and don't be shy to make written remarks and notes in them. These will often prove to be of use later for the rapid identification of the same or similar species.

When studying grasses and other plants, it is useful to collect and press specimens for later reference, but do not pick flowers as they will later bear the seeds vital for propagation. Colour photographs are a less cumbersome but more expensive way of building up your own catalogue of species.

After all is said and done, however, there is simply no substitute for experience. Only after repeated observation will you learn to interpret the signs, notice the subtle movements, and predict the whereabouts and behaviour of certain creatures.

HABITAT DESCRIPTIONS

Planet Earth may be divided into several broad categories of land type. Geographers refer to these as **vegetation zones**, and they include such well-known types as forest, grassland and desert. Recognising the interrelatedness of all life forms, ecologists now prefer to use the word **biome** for these broad definitions, so as to embrace all the living and non-living components. Generally speaking, particular plants and animals are characteristic of, and confined to, a particular biome, but there are notable exceptions, particularly among birds and larger mammals.

A biome is determined by **geology** and **climate** and although humans may modify or even destroy the landscape, its classification does not change. Within each biome, various factors – such as local rainfall, soil type and aspect – give rise to the creation of **habitats** and these may be completely altered or destroyed by human activities, which can readily turn a wooded valley into a golf course or a marsh into a dam. Such artificial environments – to which some species readily adapt – are habitats in their own right, albeit man-made.

The Johannesburg area falls within the **grassveld biome** but has a wide diversity of natural and man-made habitats. The former, regrettably, are now few in number and small in size, but still contain representative flora and fauna. On the pages that follow, seven distinct habitats are identified, their characteristics described, and some typical plants and animals highlighted. The coded symbols at the top of each habitat page are used throughout the species accounts as a means of linking the various animals and plants to their preferred habitats.

Key to habitat symbols

G   Gardens

PS   Parks and Sports Fields

WR   Wetlands and Rivers

NG   Natural Grasslands

KC   Koppies and Cliffs

S   Savanna and Scrub

RF   Riverine Forest

Gardens

Suburban gardens, of all shapes and sizes, occupy a large proportion of the Johannesburg environment. Whether living on a newly established plot, or an older suburb with large trees, a variety of wildlife can be encouraged to share your garden.

A certain amount of planning or restructuring of your garden can produce surprising results. The planting of indigenous trees and shrubs will attract insects which in turn provide food for birds, bats, lizards and frogs, as well as affording plenty of enjoyment in their own right. A tangled, bushy thicket in a corner of your garden will provide a refuge for shyer creatures, and some birds may nest there. Fallen leaves should be left in flower beds, as they not only protect the soil from drying out, but also provide homes and feeding opportunities for many species. Although all South African plants are considered indigenous to the sub-continent, many of the species from warmer climates, or from the winter rainfall zone of the south-western Cape, may not survive Johannesburg's cold and dry winters. The growth of indigenous plants is often curtailed by insects, but if certain plants cannot survive this feeding pressure it is better to let them die, and replace them with something else, than to resort to pesticides which have a domino effect on garden food chains.

The building of a small pond or, better still, the creation of a mini-wetland will entice a host of water-loving creatures, and can be stocked with indigenous fish such as the Banded Tilapia.

Providing feeding opportunities and nesting sites for birds is a most rewarding activity, and may become an intensive hobby. A useful book on this topic is *Attracting Birds to your Garden in Southern Africa* by Roy Trendler and Lex Hes (Struik, 1994).

13

Parks and Sports Fields

The broad definition for a park or sports field is an open area of short, regularly mown grass, usually dotted with tall trees but largely devoid of thickets and rank undergrowth. Public parks are regrettably few in number in Johannesburg, but include Delta Park, Zoo Lake and Emmarentia Dam. These three parks all incorporate, or surround, fairly large bodies of open water – a feature common to many of these man-made habitats.

Trees typical of Johannesburg parks are aliens such as the Weeping Willow *Salix babylonica*, American Plane *Platanus occidentalis,* Tipu Tree *Tipuana tipu* and Jacaranda *Jacaranda mimosifolia*, but indigenous species such as Wild Olive *Olea europea,* White Stinkwood *Celtis africana* and River Bushwillow *Combretum erythrophyllum* are being planted to an increasing extent.

No larger mammals are present in these parks due to the presence of people and dogs and the modified landscape which provides few hiding places. Feral Cats are common in many parks, but of indigenous mammals only nocturnal rodents and bats can be found with any certainty. A number of birds have adapted readily to these open areas, with Hadeda Ibis, Cattle Egret, Crowned Plover, Spotted Dikkop and Egyptian Goose being common in the larger parks. Crested Barbet and Redthroated Wryneck often excavate nesting holes in the soft wood of trees such as the Weeping Willow and Match Poplar *Populus deltoides*.

These same species may also be found on school and club sports fields and on golf courses. Golf courses are particularly good places for birds and other wildlife – as they often incorporate dams and streams – but access is limited strictly to members.

Wetlands and Rivers

A number of streams drain to the north and south of the Witwatersrand ridge but the suitability of these for wildlife is dependent upon pollution levels, and much of the natural vegetation of the banks has been removed by local authorities. Today, stretches of the larger perennial streams such as the Braamfontein Spruit, Sand Spruit and Bloubos Spruit have been incorporated into parks, and act as corridors for wildlife and nature enthusiasts alike.

On the quieter and least disturbed stretches of these streams, and often in places where small dams have been created, kingfishers, ducks and the more hardy fish species may be found.

Much more attractive to aquatic wildlife are the wetlands on the East and West Rand where the Common Reed *Phragmites australis*, Bulrush *Typha capensis* and various sedges and grasses are the dominant plants. The Natal Spruit winds its way through the industrial areas of Springs, Germiston and Alrode, and although effluent does enter these waters, much of it is rich in nitrogen and actually increases the density of small lifeforms. Highly toxic chemicals have been found leaking into this system, however, with devastating effect on fish and other aquatic wildlife.

Bird sanctuaries such as Rondebult and Marievale are situated on the Natal Spruit and offer outstanding birdwatching experiences – flamingos, herons, ducks, grebes, ibis and waders are among the birds regularly seen. Less extensive wetlands occur on the West Rand, where the Con Joubert Sanctuary near Randfontein is an important waterfowl refuge. South of the Klipriviersberg, the floodplain of the Klip River abounds with fish, frogs and reed-loving birds.

Natural Grasslands

Once covering much of the area, natural grasslands are now very scarce in the vicinity of Johannesburg. Typified by undulating, open country devoid of woody plants, Highveld grasslands include a wide diversity of grass species, with Rooigras *Themeda triandra* being one of the most dominant species. A host of annual flowering plants, including many lilies, are also characteristic.

Under natural conditions, grasslands are maintained by fire which, through the repeated burning down of tree and shrub saplings, prevents the otherwise inevitable encroachment of bush. Although some trees and shrubs of the Highveld are fire-resistant – Common Sugarbush *Protea caffra* and Mountain Cabbage Tree *Cussonia paniculata* are two examples – this only applies once they have reached a certain age and developed a thick layer of bark. Grasslands may be damaged by too frequent burning, however, as some grass species may be displaced by other plants. Accidental fires are a regular and repetitive event in the area. When grassland soils are disturbed by man's activities, 'pioneer' plants, including well-known aliens such as Khakiweed *Tagetes minuta* and Blackjack *Bidens pilosa* take hold. Roadsides are favourable sites for colonisation by these and the pretty Cosmos *Bidens formosa* which flowers along Johannesburg's country roads in autumn.

Fairly extensive open grasslands are to be found in the Krugersdorp Game Reserve, where a variety of large herbivorous mammals once common in the area have been reintroduced. Less extensive grasslands exist north of Fourways. Also of interest are the hilltop grasslands of the Klipriviersberg Nature Reserve and the marshy grasslands in the Marievale Bird Sanctuary.

Koppies and Cliffs

The single biggest rock formation of the area is the Witwatersrand ridge which forms a low escarpment stretching from Orange Grove in the east to beyond Roodepoort in the west. This ridge supports an interesting and, in some places, still relatively intact flora, including plants such as the Redleaved Rock Fig *Ficus ingens* and Cape Gardenia *Rothmannia capensis*. In some places, such as on Northcliff Hill and Witpoortjie Falls, exposed cliffs occur and are home to Rock Dassies and, at the latter locality, a pair of Black Eagles. Mostly out of reach of veld fires, these cliffs support some ancient trees – such as Wild Olive *Olea europea* and Jacket Plum *Pappea capensis* – which may have been alive when gold was first discovered here in 1886. In recent times town planners have unfortunately not restricted the development of residential properties on the crest of the ridge, and obtrusive structures have spoilt much of the skyline.

Many of the exposed rocks in the Klipriviersberg south of the Witwatersrand are hardened lavas of volcanic origin, and have created an attractive landscape of reddish boulders and koppies. Rock Dassie and Jameson's Red Rock Rabbit are found here alongside birds such as Mountain Chat, Rock Bunting and Cape Rock Thrush. Reptiles such as the Ground Agama and Striped Skink are plentiful. Rock-adapted plants such as Mountain Aloe *Aloe marlothii* and Transvaal Milkplum *Englerophytum magalismontanum* are conspicuous.

One of the most interesting rock features in the area is Lone Hill, a jumble of finely balanced granite boulders north of Bryanston. Rocky habitats are also to be found in suburbs such as Northcliff, Linksfield, Melville, Randpark Ridge and Bryanston.

Savanna and Scrub

Areas of savanna are found in various localities, most extensivly in the Klipriviersberg, Melville Koppies and along the western end of the Witwatersrand ridge. Two distinct forms are present: **acacia savanna**, dominated by Common Hookthorn *Acacia caffra* and usually growing on north-facing slopes; and **protea savanna**, dominated by Common Sugarbush *Protea caffra* and typical of south-facing slopes. These dominant trees, along with numerous others, are usually of no great height and the crowns of adjacent trees rarely touch. Various grasses, including Guinea Grass *Panicum maximum,* grow beneath the mostly deciduous trees. Runaway fires keep savanna from becoming scrubby by removing fire-sensitive saplings and shrubs.

Birds are particularly common in acacia savanna, with Acacia Pied Barbet, Whitebellied Sunbird and Redfaced Mousebird being among the residents. Protea savanna is favoured by Redthroated Wryneck, Malachite Sunbird, Streakyheaded Canary and others. Smaller mammals and reptiles may be fairly common in either savanna type.

In disturbed or denuded areas, where the absence of grass prevents seasonal fires, **scrub** – consisting of low-growing Common Hookthorn *Acacia caffra*, Common Wild Currant *Rhus pyroides*, and Common Spikethorn *Maytenus heterophylla* – often occurs. Invasive alien plants such as Black Wattle *Acacia mearnsii* and Sweet Prickly Pear *Opuntia ficus-indica* may flourish under such conditions.

At the base of koppies and cliffs, and along streams, trees grow taller and closer together, with species such as White Stinkwood *Celtis africana,* Wild Pear *Dombeya rotundifolia* and Wild Peach *Kiggelaria africana* forming thickets.

Riverine Forest

Riverine forest is typified by tall trees with interlocking crowns, and a tangled understorey dominated by shrubs and creepers. Grasses are sparse or absent. The ready availability of ground water allows large trees to grow close together, but they must then compete for light and this results in tall trunks and relatively small crowns.

Only where stream banks have been left undisturbed do fragments of this species-rich habitat remain in Johannesburg. Perhaps the best remaining example is to be found along the Witpoortjie River in the Witwatersrand Botanical Garden, where large White Stinkwood *Celtis africana* and River Bushwillow *Combretum erythrophyllum* are the dominant trees. North of Fourways, stretches of the Jukskei River are still bordered by mature riverine forest. Sweet Thorn *Acacia karroo* is characteristic of the outer fringe of this habitat, while Sagewood *Buddleja salviifolia* and Oldwood *Leucosidea sericea* crowd the river banks. The Royal Paintbrush Lily *Scadoxis puniceus*, which blooms in spring, is among the beautiful smaller plants. Typical birds include the Redchested Cuckoo, Paradise Flycatcher and Southern Boubou, and butterflies are particularly common.

Where streams run through suburbia, authorities have mostly cleared the natural vegetation and alien Weeping Willow *Salix babylonica* and Kikuyu Grass *Pennisetum clandestinum* prevail. Such areas are usually kept open with mowers and chainsaws, and fall into the *Parks and Sport Fields* habitat type (see p. 14). Invasive alien species such as Black Wattle *Acacia mearnsii*, Grey Poplar *Populus canescens*, Bugweed *Solanum mauritianum* and Brazilian Glory Pea *Sesbania punicea* invariably take over when indigenous species are removed.

Mammals

In comparison to other parts of South Africa, the Johannesburg area has an impoverished mammal fauna with only the smaller, more secretive and nocturnal species being able to occur in such close proximity to man.

Featured in this section are the mammals which occur commonly in the area as well as some which, although rarely seen, are of great interest and should be looked out for. Only a few of the many rodents and bats are included.

In years gone by, large herds of grazing herbivores, including White Rhino, were probably quite common, while Kudu and other browsers no doubt occurred on wooded hillsides. Needless to say, this impressive array of animals has long since disappeared from the area, although small numbers have been reintroduced into the Krugersdorp Game Reserve and elsewhere.

In recent years there have been extraordinary sightings of Leopard and Brown Hyena on the outskirts of suburbia, but today's mammal-watchers will have to content themselves with smaller, but often no less interesting, creatures. Getting out and about in the early hours, and being aware of tracks and signs, offers the best opportunities to glimpse the smaller mammals.

Names used in the following accounts follow those in the standard reference work – *The Mammals of the Southern African Subregion* by J. Skinner and R. Smithers (Univ. of Pretoria, 1990). Chris and Tilde Stuart's *Field Guide to the Mammals of Southern Africa* (Struik, 1988) is a less detailed but more portable reference book. Assistance with the identification of various rodents and bats may be obtained from the Transvaal Museum in Pretoria.

Hedgehog

Small nocturnal insectivore with **short black and white spines**, and a pointed snout. Closely related to shrews, it is a voracious predator of termites, worms, crickets and other small creatures. Days are spent in a burrow or among dense vegetation. If disturbed it rolls itself into a ball. Hibernates during midwinter. Useful in the garden, they are often caught or harassed by domestic dogs.
Length: 20 cm Mass: 400 g

Porcupine

Very large nocturnal rodent with **long black and white quills**. Vegetarian, it is an avid digger, feeding primarily on roots and tubers. Tree bark is also favoured. Days are spent in a burrow or among rocks. Its presence may be detected by gnaw marks on the trunks of trees, or discarded quills. When threatened, the quills may be raised and rattled.
Length: 75 to 100 cm Mass: 10 to 24 kg

Rock Dassie

Compact, short-eared mammal with **no obvious tail**. Active by day, it frequently basks in the sun, particularly in early mornings and on cool winter days. Colonies comprise several family units. Latrines are characterised by white and brown urine streaks on rocks, and **spherical pellets**. The diet includes grass, leaves and berries. May climb into trees to feed. Calls with a sharp bark.
Length: 50 cm Mass: 4 to 5 kg

MARK TENNANT/AFRICAN IMAGES

Cape Hare

Long-legged and long-eared mammal which **favours undisturbed grassland**. Very difficult to distinguish from the Scrub Hare, but slightly smaller and often has an **indistinct buffy band** on its side, between the off-white underbelly and fawn-grey back. Mostly nocturnal, it feeds on grasses and herbs. May be flushed by day from its hiding place among grass, running off in a zig-zag manner.
Length: 48 cm Mass: 2 kg

NATIONAL PARKS BOARD OF S.A.

Scrub Hare

Long-legged and long-eared mammal which **favours savanna and scrub**. Very difficult to distinguish from the Cape Hare, but slightly larger and has no **buffy band** on its side. The underbelly is off-white and the back fawn-grey. Mostly nocturnal, it feeds on grasses and herbs. May be flushed by day from its hiding place among low bushes or grass tufts, running off in a zig-zag manner.
Length: 54 cm Mass: 2 to 3 kg

Jameson's Red Rock Rabbit

Long-eared mammal which **favours rocky outcrops**. May be distinguished from the similar hares by its more compact build, rich **russet-red coat**; and somewhat shorter ears. Mostly nocturnal, it feeds on grasses and leaves. May be flushed by day from its hiding place among rocks. **Lozenge-shaped pellets** are deposited in urine-free latrines among grass.
Length: 45 cm Mass: 2 to 3 kg

NATIONAL PARKS BOARD OF S.A.

Slender Mongoose

Small carnivore with **reddish coat** and **very long black-tipped tail**. Often seen crossing roads with its tail held aloft. Occurs singly or in pairs in a variety of habitats, including gardens adjacent to rocky outcrops and parks. Most of its time is spent on the ground, but this mongoose is also an adept tree climber. Its diet includes insects, birds' eggs and lizards. Active mostly by day.
Length: 60 cm (incl. 27 cm tail)   Mass: 650 g

Yellow Mongoose

Small carnivore with pale **yellowish coat** and long **white-tipped tail**. Prefers more open habitats than the Slender Mongoose, and is frequently encountered in the vicinity of bare pans and vleis. Its diet includes beetles, lizards and birds' eggs. Gregarious by nature, family groups of up to ten members excavate burrow systems in open areas. Active only by day.
Length: 60 cm (incl. 24 cm tail)   Mass: 830 g

Smallspotted Genet

Small carnivore with boldly patterned **black and white coat and very long bushy tail**. Occurs singly in wooded or rocky habitats, but may live in larger gardens. An agile climber, it spends most of its time in trees. Its diet includes rodents, reptiles and nestling birds. Large, cigar-shaped droppings accumulate at conspicuous latrine sites. Nocturnal and secretive.
Length: 95 cm (incl. 45 cm tail)   Mass: 2 kg

MANFRED REICHARDT

Feral Cat *

Small, highly adaptable carnivore with variable coat colour and pattern. Tens of thousands of stray, escaped and unwanted cats frequent alleys, parking lots and drains. Breeds prolifically. Most active at night; rodents, birds and scavenged pickings provide the bulk of the diet. Where they occur in nature reserves and parks, they pose a threat to indigenous fauna, including smaller carnivores.
Length: up to 90 cm   Mass: up to 5 kg

Blackbacked Jackal

Dog-like carnivore with **fawn and black coat and long bushy tail**. Rarely seen, its haunting wail may still be heard after dark on the outskirts of suburbia. Strictly nocturnal, it feeds upon a very wide range of food items including rodents, hares, insects, lizards, birds and carrion. Occurs in pairs or family groups. Pups are born in an underground burrow during late winter or spring.

Length: 1 m (incl. 32 cm tail) Mass: 8 kg

Rock Elephant Shrew

Very small insectivore with **long, tube-like snout**, large rounded ears and sandy coat. Occurs singly or in family groups in rocky areas. Active during daylight hours, particularly in early morning and evening, when it dashes to and from cover. Beetles, termites and other insects are the main prey. Retreats into its den at midday and at night. The name is derived from its elongated snout.

Length: 26 cm (incl. 13 cm tail) Mass: 60 g

C & T STUART

Reddish-grey Musk Shrew

Miniscule insectivore with **long snout**, small rounded ears and rust-grey coat. Occurs singly or in family groups. Active at night, they may hunt for insects around outdoor lights. Despite their size, shrews are voracious predators, able to eat their own body weight in food each night. Often caught by domestic cats or drowned in swimming pools. Several similar species are found in the area.

Length: 10 cm (incl. 4 cm tail) Mass: 8 g

NATIONAL PARKS BOARD OF S.A.

Lesser Bushbaby

Very small tree-dwelling primate with **huge eyes, long bushy tail** and grey-brown coat. Occurs in pairs or family groups. Active at night, they leap between branches, following demarcated trails to favoured feeding sites. Insects, birds' eggs, berries and the sweet resin of acacia trees are featured in the diet. Roosts by day, and rears young in a tree hole or nesting-box.

Length: 38 cm (incl. 22 cm tail) Mass: 150 g

LEX HES

Common Duiker

Small antelope with a grey-brown coat. The stocky build, **narrow ears** and **dark band down the centre of the face** are diagnostic. Both sexes have a tuft of long hair on the crown but only the male has a pair of short, ringed horns. Forages singly or in pairs, mostly after dark. Browses on leaves. Small numbers occur in the Klipriviersberg and perhaps elsewhere.

Length: 1 m Height: 60 cm Mass: 19 kg

Steenbok

Small, slender antelope with a rufous coat. The **large rounded ears** and long legs distinguish it from the larger Common Duiker. Only the male has a pair of sharply-pointed horns. Prefers open grassland to scrub or bush. Forages in pairs, sometimes after dark, feeding on leaves and grass. Small numbers occur north of Fourways and elsewhere on the outskirts of the region.

Length: 85 cm Height: 50 cm Mass: 11 kg

Klipspringer

Small, stocky antelope with grizzled, grey-brown coat. The **prominent orbital glands** and **habit of walking and jumping on hoof tips** are diagnostic. Only the male has a pair of short, pointed horns. Occurs in pairs or family groups on rocky outcrops or cliffs, where it feeds on leaves of small shrubs and herbs. Small numbers occur in the Klipriviersberg and perhaps elsewhere.

Length: 85 cm Height: 60 cm Mass: 13 kg

Grey Rhebok

Medium-sized antelope and the largest mammal occurring naturally in the area. The **woolly coat, mule-like ears** and **thin pointed horns** of the male are diagnostic. Occurs in family groups on grassy hillsides, where it grazes on a variety of grasses. Utters a sharp snort if disturbed and runs off with a rocking gait. Seen only on a sporadic basis in the Klipriviersberg.

Length: 135 cm Height: 75 cm Mass: 20 kg

Cape Serotine Bat

Very small, insectivorous bat with pale brown back and off-white underparts. The **snout is pointed and the ears short**. This is the little bat most often seen flying at dusk in the company of swallows and swifts. Small airborne insects such as mosquitoes and moths are hunted well into the night. By day, it roosts in colonies under the eaves of houses or in tree holes.

Length: 8 cm Wingspan: 23 cm Mass: 6 g

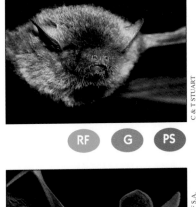

C & T STUART

RF G PS

Common Slitfaced Bat

Small, insectivorous bat with pale brown back and pure white underparts. The **very long, rounded ears** are obvious. The wings are rounded and the tail membrane has a notched tip. A long slit runs down the centre of the face, but is inconspicuous. Roosts in colonies of several hundred in caves or man-made structures. Rarely emerges until well after dark.

Length: 10 cm Wingspan: 24 cm Mass: 11 g

NATIONAL PARKS BOARD OF S.A.

PS G KC

Geoffroy's Horseshoe Bat

Small, insectivorous bat with fawn or ginger body. The ears are large and pointed. The tail membrane has a pointed tip. Roosts in colonies of up to 1 000 or more in caves, mine adits and other structures. The **elaborate 'nose-leaves'** – not likely a horseshoe in structure – are diagnostic. Emerges from its roost at dusk, hunting small insects throughout the night.

Length: 10 cm Wingspan: 32 cm Mass: 17 g

G PS KC

Yellow House Bat

Medium-sized, insectivorous bat, with distinctive **yellow belly** and darker head and upper-parts. The **snout is dog-like**. The tail membrane has an acutely pointed tip. Partial to built-up areas, small numbers roost under eaves and other overhangs. Emerges after dark to forage around outdoor lights and may even enter houses in pursuit of flying insects.

Length: 13 cm Wingspan: 30 cm Mass: 27 g

NATIONAL PARKS BOARD OF S.A.

PS G

Common Molerat

Small, short-legged rodent with **tiny eyes, hidden ears**, and **protruding teeth**. This is the creature responsible for the mounds of earth which appear on lawns and sports fields. Abundant, but rarely seen due to its underground lifestyle. It spends the daylight hours below ground, where it burrows through the soil in search of roots and bulbs. It may emerge from its burrow system at night.
Length: 15 cm Mass: up to 150 g

PS G NG

House Mouse *

Small, long-tailed rodent with **large rounded ears** and grey-brown coat. Nocturnal and omnivorous. Dependent upon the dwellings and refuse of humans. Untidy nests are made of paper and other rubbish. Breeds prolifically. Thought to have originated in northern Europe, local populations probably expanded from ports such as Cape Town and Durban.
Length: 16 cm (incl. 9 cm tail) Mass: 18 g

G PS

Multimammate Mouse

Fairly small, long-tailed rodent with **large rounded ears** and grey-brown coat. Larger than the similar House Mouse. Nocturnal and omnivorous. May occur in houses and buildings. Its name refers to the extraordinary number of teats – up to 12 pairs – possessed by the female, a factor which allows this to be the most rapidly reproducing mammal in Africa. Population explosions may occur.
Length: 24 cm (incl. 11 cm tail) Mass: 60 g

NG S G

Striped Mouse

Fairly small, long-tailed rodent with rounded ears and **four distinct stripes** running down its back. Unlike most other mice, it is most active during the day. It rarely, if ever, enters households as open grassland is the preferred habitat. Feeds mostly on grass seeds. Extremely common, it is the favoured prey of the Blackshouldered Kite, Greater Kestrel and Marsh Owl.
Length: 20 cm (incl. 9 cm tail) Mass: 45 g

NG S

Angoni Vlei Rat

Medium-sized, **short-tailed** rodent with a **blunt snout** and rounded ears. The dark coat is shaggy. Occurs in dense vegetation near water, where it is active during both night and day. It makes use of runs and tunnels in order to reach feeding sites. Grass roots, and the new shoots of reeds and sedges are favoured. Marsh Owl and African Marsh Harrier are among the many predators of this abundant rodent.
Length: 30 cm (incl. 8 cm tail) Mass: up to 250 g

C & T STUART

WR

Woodland Dormouse

Small, silvery-grey rodent with pale underparts. The long and **bushy, squirrel-like tail** is diagnostic. Strictly nocturnal, it spends most of its time in trees, particularly thorny acacias. It may also live in buildings and structures, and sometimes enters houses. Its diet includes insects, spiders and seeds. A substantial nest of grass, lichen, leaves or paper is built within a tree hole or similar cavity.
Length: 16 cm (incl. 7 cm tail) Mass: 30 g

NATIONAL PARKS BOARD OF S.A.

KC G S

House Rat *

Large, long-tailed rodent with rounded ears and grey-black coat. Its completely **naked tail** is diagnostic. It lives in close proximity to man, particularly in warehouses and storerooms where untidy nests are made with a variety of odds and ends. Like the House Mouse, this species arrived first at South Africa's ports before spreading inland to major towns. Omnivorous and destructive.
Length: 37 cm (incl. 20 cm tail) Mass: 150 g

NATIONAL PARKS BOARD OF S.A.

PS G

Highveld Gerbil

Fairly small, long-tailed rodent with reddish-grey coat. The **tail is usually white tipped**. Strictly nocturnal, it bounds about on its hind feet in search of food or to escape predators. Grass seeds, berries and insects make up the diet. Rarely seen but its presence may be detected from the conspicuous colonies – a number of small mounds and little holes in sandy soil.
Length: 28 cm (incl. 14 cm tail) Mass: 80 g

C & T STUART

S NG

Birds

Over 300 species of birds have been recorded in the Johannesburg area, although many of these are only sporadic visitors. In this section 142 of the more common and interesting species are described and illustrated. Some birds are confined to particular areas, but are included on the basis that they may be fairly predictably found if a specific locality is visited.

Birdwatching is most rewarding during the summer months when many species are singing and breeding, and when numerous migrants arrive from further afield. During the winter months, however, more birds move into suburban gardens as surrounding natural habitats dry out. Leafless trees permit easier observation of some smaller birds in winter.

The species are arranged in such a way that those which could be confused appear on the same spread, and this has resulted in a slightly different sequence from that of the standard reference works. The names used are those recognised by the South African Ornithological Society (SAOS). Two comprehensive field guides are available in the form of *Newman's Birds of Southern Africa* (Southern, 1992) and *Sasol Birds of Southern Africa* (Struik, 1993); any serious birdwatcher should own at least one of these books. A list of more general publications is provided on p. 122. There are three bird clubs in the area (addresses on p. 123), and joining one of these will provide you with opportunities to meet experienced birdwatchers, visit localities often closed to the general public, participate in conservation projects, and receive popular and scientific periodicals.

Dabchick (Little Grebe)

Small waterbird, superficially resembling a duck. The **chestnut neck and pale spot at the base of the bill** – of breeding adults – are diagnostic, as is the habit of repeatedly diving below the surface of the water. Small fish, crabs and frogs are caught underwater. May be found on any stretch of open water, but favours pans. A loud trill is the call, often uttered during courtship.
Length: 20 cm

MANFRED REICHARDT

Great Crested Grebe

Medium-sized waterbird, superficially resembling a duck. The **long white neck** and **peaked crest of black feathers** are diagnostic. It favours larger dams and pans, keeping to deepest water. Fish are caught underwater. The courtship ritual involves both partners raising themselves out of the water and stretching out their necks. Most common in the wetlands of the East Rand.
Length: 50 cm ss: Blacknecked Grebe

28

Reed Cormorant

Medium-sized, black waterbird with long tail and **red eyes**. Immature birds have an off-white breast. In common with other cormorants, the feathers are not waterproof and it often perches with its wings outstretched so that they may dry off. Frequent on dams, less often on streams. Fish are caught underwater. Breeds colonially in large trees, often in the company of herons.
Length: 52 cm

Whitebreasted Cormorant

Large black waterbird with a **snow-white throat and chest**, patch of bare yellow skin at the base of the bill and **green eyes**. Immatures are dark brown with completely white underparts. Less confiding than its smaller relative, it is quick to take flight. Large fish are the favoured prey and are caught underwater. Bulky stick nests are normally built in the branches of a dead tree.
Length: 90 cm

Darter

Large, dark brown waterbird with a very long neck – often held in an S-bend shape – and a sharp, pointed bill. Adults have a **rufous throat edged in white** when breeding. The wings are held out-stretched to dry. Dives for fish, which are speared before being brought to the surface. Often swims with only the neck and head above water, leading to its alternative name of Snakebird.
Length: 80 cm

Hamerkop

Medium-sized, dull brown bird, with unique **backward-pointing crest and pointed bill** giving the head a hammer-like shape. Waits quietly at the water's edge to catch frogs and other aquatic life. May be found at any stretch of water in the area, and sometimes visits fish ponds in gardens. A huge dome-shaped nest of twigs and mud is built in the branches of a large tree.
Length: 55 cm

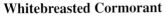

IAN SUTHERLAND/AFRICAN IMAGES

PS NG WR

Cattle Egret

Small, all-white egret with a relatively short neck. The legs and bill are yellowish-green but turn to coral-pink in the summer breeding season, when buffy plumes adorn the head and mantle. Forages on **dry land**, often in association with tractors or grazing cattle which disturb insect prey. Breeds and roosts near water, sometimes in the company of other birds.
Length: 55 cm ss: Little Egret (65 cm); Great White Egret (95 cm); Squacco Heron (42 cm)

LEX HES

WR

Grey Heron

Large, pale grey heron with **white head and neck**, and long **yellow bill**. A bold black streak runs above and behind the eye to form a small crest. In flight, the **underwing is uniform grey**. Flies with neck tucked in. Usually seen alone at the edge of dams, pans and streams, it feeds on frogs, fish and other aquatic animals. The nest is a platform of reeds and sticks, built in a tree or reedbed.
Length: 100 cm ss: Goliath Heron; Purple Heron

MANFRED REICHARDT

NG WR

Blackheaded Heron

Large, dark grey heron with **black head and neck**, and long **grey bill**. The throat is white. In flight, the **underwing is black and white**. Flies with neck tucked in. The immature bird is paler with a yellow lower bill. Feeds away from water in short grassland, farmlands, and along roadsides where it catches rodents and large insects. Nests are near water, in overhanging trees or reedbeds.
Length: 95 cm ss: Blackcrowned Night Heron

WR

African Spoonbill

Large, all-white waterbird with distinctive spoon-shaped bill. The bare face of the adult is pink. Flies with neck out-stretched. The unusual bill is used in a sweeping motion in shallow water to capture small crustaceans and aquatic insects; small fish and frogs may also be taken. Often nests in the company of other species – particularly Sacred Ibis – in reedbeds or in large trees.
Length: 90 cm

Hadeda Ibis

Stocky, olive-brown ibis with **short legs** and long decurved bill. Adults have a reddish upper bill and a metallic sheen of violet and emerald on the shoulders. The call is a raucous 'ha-de-dah', often uttered at dawn and dusk. Has adapted well to modified landscape. Pairs or small flocks are a familiar sight on garden lawns and in parks. A stick nest is built within the canopy of a leafy tree.
Length: 75 cm

RF   G   PS

Glossy Ibis

Slender, chocolate-brown ibis with **long legs** and long decurved bill. Adults have a metallic bronze sheen and a white stripe above the eye. This elegant bird is always found near water, where it wades in the shallows in search of frogs and aquatic insects. Occurs singly or in flocks of up to 50. Platform nests are made in reedbeds. Flocks fly in graceful V-formation to and from their roosts.
Length: 70 cm

WR

Sacred Ibis

Slender white ibis with long legs and long, scythe-shaped bill. The **naked black head and neck** are diagnostic. Breeding adults have the tail adorned with black plumes. The flight feathers are tipped in black. The name refers to its place in Egyptian mythology. It has the rather unsavory habit of foraging around rubbish dumps. Graceful and elegant when airborne, flocks fly in V-formation.
Length: 90 cm

PS   WR

Greater Flamingo

Slender white bird with extremely long neck and pink legs. The **bill is pale pink with a dark tip** (the smaller Lesser Flamingo has a dark maroon bill). The outstretched wings are pink and black and spectacular in flight. Immature birds are dusky white with a pale bill and grey legs. Occurs in small flocks, feeding on aquatic micro fauna in shallow water. Frequent on East Rand pans.
Length: 140 cm ss: Lesser Flamingo (100 cm)

WR

31

Egyptian Goose

Large, fawn and chestnut goose with **pale face and brown eye mask**. In flight, the **white, green and black wings** are conspicuous. Occurs in pairs or family groups in open areas where it feeds on grass, grain and bulbs. Sometimes perches on street lamps. Very noisy and bold but wary of people, it may become pugnacious during the breeding season. Population is on the increase.

Length: 70 cm ss: South African Shelduck

Spurwinged Goose

Very large, green-black goose with a variable amount of **white on the face, wings and underbelly**. The bill and legs are pink. Usually occurs in flocks which spend the day resting on mudflats or in shallow water, and fly off to feeding grounds in the evening. Grass, grain and tubers comprise the diet. Call is a muted whistle. The male is much larger than the female.

Length: 100 cm (male)

African Black Duck

Medium-sized, charcoal-black duck with **back spotted in white, and orange legs**. In flight, the **sapphire-blue wing panels** rimmed in white are conspicuous. Unobtrusive, it **prefers the moving water of streams and rivers**. Tubers, seeds and insects are eaten. Shy and wary, it keeps to vegetated banks, but may rest on rocks. Pairs defend territories along streams. The call is a loud 'quack'.

Length: 56 cm

Whitefaced Duck

Medium-sized, long-necked duck with an **upright stance**. The **white face**, chestnut breast, and black and white barred flanks are diagnostic. The call is a lovely, flute-like whistle made in flight or at rest. Very gregarious, it may gather in flocks of a hundred or more at pans and lakes. Spends much of its time resting in flocks on mudflats at the water's edge. Feeds on aquatic tubers and seeds.

Length: 48 cm ss: Fulvous Duck

Yellowbilled Duck

Medium-sized duck with a **bright yellow bill**. The grey back feathers are edged in white to give a scaly appearance. In flight, the **emerald green wing panels**, rimmed in white, are conspicuous. Occurs on small dams and larger expanses of water. May join domestic ducks in parks to feed on bread, but at such times it is likely to come into contact with the next species with which it may interbreed.
Length: 57 cm

Mallard Duck *

Medium-sized duck with sexes differing in appearance. The breeding male has an **emerald-green head**, yellow bill and chestnut breast; the female and non-breeding male is mottled brown with a cream eyebrow and dark bill. An **alien species** which has escaped from captivity to form free-flying populations. Poses a threat to the genetic purity of the Yellowbilled Duck.
Length: 58 cm

PS WR

Cape Shoveller

Medium-sized duck with long flattened bill. The body is pale grey-brown with a speckled appearance. Males have pale heads, **lemon-yellow eyes** and **bright orange legs and feet**. Females have dark eyes and feet. Males frequently chase females across the water. In flight, the turquoise wings and green wing panels are conspicuous. Often seen in the company of other ducks.
Length: 53 cm

Hottentot Teal

Very small duck with dark mottled plumage, **black cap** and **sky-blue bill**. The underparts are pale fawn. In flight, the white tips to the secondary flight feathers are conspicuous. Unobtrusive, it usually keeps within the fringing vegetation of pans and lakes but often allows a close approach. Feeds on seeds, floating vegetation and insects.
Length: 35 cm

Redbilled Teal

Small duck with dark mottled plumage, **black cap** and **crimson-red bill**. The underparts are pale fawn. In flight, the fawn and white wing panel is conspicuous. Usually swims among vegetation. May gather in large flocks of several hundred. Prefers shallow water and is quick to arrive at flooded grasslands. Food consists of grass seeds, grain and small aquatic creatures.
Length: 48 cm

Cape Teal

Small duck with pale mottled plumage, **coral-pink bill** and **red eyes**. It differs from the darker Redbilled Teal in lacking a black cap. In flight, the white panel with a green band is conspicuous. Prefers stretches of open water and is usually seen in pairs or small family groups. Often associates with other ducks. It may dive below the surface for aquatic insects and bulbs.
Length: 46 cm

Southern Pochard

Small, maroon-brown duck. The male is very dark with a pale, **grey-blue bill** and **bright red eyes**. The female is paler with a white, C-shaped line on the neck, and dark eyes. In flight, the long white wing bar is conspicuous. Usually occurs in pairs or small groups, often in the company of other ducks. May be rare or absent during the winter months. Often reluctant to leave the water.
Length: 50 cm

Maccoa Duck

Small, chestnut-brown duck which **sits low on the water** with stiff tail feathers held erect or fanned-out on the surface. The male has a satin-black head and a **bright blue bill**. The female is darker with a pale throat and a thin, off-white streak below the eye. Repeatedly dives for food which includes tubers, seeds and small aquatic creatures.
Length: 46 cm

Redknobbed Coot

Black waterbird with **white bill and frontal shield**, complemented by red eyes and a pair of red 'knobs' on the forehead. It is one of the commonest waterfowl and may congregate in large numbers when not breeding. The nest of reeds is built on open water. Pugnacious and aggressive, waterborne chases between rivals are commonplace. Feeds on aquatic plants.
Length: 44 cm

WR

Moorhen

Black waterbird with **red frontal shield and bill with a yellow tip**. The legs and feet are bright yellow and there is a thin white streak running down each flank. Smaller, and with a lighter build than the previous species with which it often shares the same habitat. When out of the water, the short tail is repeatedly flicked up and down to reveal snow-white under-feathers. Feeds on aquatic plants.
Length: 33 cm

LEX HES

WR

Purple Gallinule

Purple-green waterbird with **coral-pink frontal shield, bill and legs**. The back is olive-green and the under-tail snow-white. Secretive and seldom seen, it keeps to reedbeds and dense vegetation. Occasionally it may stride out from cover on its long legs to forage in shallow water or on mudflats, but retreats quickly if disturbed. The diet includes tubers, sedge stems and nestling birds.
Length: 45 cm

WR

Black Crake

Small black waterbird with **lime-yellow bill** and **red legs and eyes**. Secretive and seldom seen, it keeps to reedbeds and dense vegetation but is nevertheless something of a 'show-off' in comparison to other crakes. May be located by its harsh throaty call – a duet between a pair. The toes are very long and allow it to run across floating vegetation. Feeds mostly on aquatic insects and worms.
Length: 21 cm ss: Baillon's Crake; African Rail

WR

Common Sandpiper

Small, grey-brown wader with fairly short, grey-green legs and thin bill. The **plain brown back**, and **white shoulder patch in the shape of an inverted C**, are diagnostic. Common **summer migrant** usually found foraging alone on mud flats, along streams or in roadside pools. When walking, the tail is constantly bobbed up and down. In flight, the long **white wingbars** are conspicuous.

Length: 19 cm ss: Little Stint; Curlew Sandpiper

Wood Sandpiper

Small, grey-brown wader with fairly long yellow legs and thin bill. The dark **back is boldly spotted in white** and the white eye-stripe extends to the back of the head. Common **summer migrant** usually found on mudflats and in shallow water. Many individuals may forage around a single body of water, but they tend to space themselves out. In flight, the **white rump** is conspicuous.

Length: 20 cm

Marsh Sandpiper

Slender, pale-grey wader with very long, grey-green legs and thin bill. The back has a scaly appearance. Common **summer migrant** usually found on mudflats and in shallow water. Very restless, this stilt-like bird is constantly on the move as it probes the mud for worms and insect larvae. In flight, the **white rump and back** are conspicuous. Considerably smaller than the similar Greenshank.

Length: 23 cm ss: Greenshank (34 cm)

Ruff

Heavy-bodied, grey-brown wader with fairly long legs and a thin bill about the same length as the head. The legs may be orange or black. The back has a scaly appearance. Common **summer migrant** often occurring in large flocks on mud-flats. It is very gregarious, and foraging flocks often take off in unison to whirl around before resettling. Males are considerably larger than females.

Length: 30 cm (male) 24 cm (female)

Blackwinged Stilt

Slender, black and white wader with extraordinarily **long red legs**, and thin pointed bill. Immature birds have grey smudges on the head. Present throughout the year but prone to seasonal movement depending upon water levels in wetlands. Spends most of its time in shallow water, regularly bending over to probe the mud for worms and insects. Often in the company of other birds.
Length: 38 cm

Avocet

Slender, black and white wader with long legs and distinctive upturned bill. Immature birds are dusky grey on the back. Present throughout the year but prone to seasonal movements depending upon water levels in wetlands. Most reliably found at Rondebult. Feeds on tiny aquatic creatures, often in the company of flamingos and migratory waders in its preferred habitat of shallow water.
Length: 43 cm

Threebanded Plover

Very small plover with plain brown back contrasting with white underparts, and **red eye-ring**. The common name is something of a misnomer as there are only **two black chest bands**. Occurs singly or in pairs on mudflats and stony stream-beds, where it moves around busily in search of small insects and worms. Often associates with migrant waders in summer.
Length: 18 cm ss: Kittlitz's Plover

Ethiopian Snipe

Cryptically-plumaged wader with **short legs** and **extremely long bill**. The body is streaked, barred and spotted in chestnut and fawn, providing superb camouflage amongst reeds and grasses. Seldom seen but not uncommon. Probes deep into mud for worms and insect larvae. At the onset of winter, pairs engage in high-speed courtship flights featuring a drumming sound created by the stiff tail feathers.
Length: 28 cm

37

BETH PETERSON/AFRICAN IMAGES

Crowned Plover

Sandy-brown plover with a pure white underbelly, and black and white cap. The long **red legs** are diagnostic. A noisy and abundant resident. Occurs in pairs or small flocks in short grassland, including golf courses and street verges. In defence of eggs or young, they will circle above intruders, dive-bombing and calling loudly, and will even feign injury in a bid to distract predators from their nest.
Length: 30 cm

PS NG

Wattled Plover

Grey-brown plover with a pair of floppy yellow wattles at the base of the yellow bill. The long **yellow legs** are diagnostic. A small white patch is visible on the crown. Very noisy, but less demonstrative than Crowned or Blacksmith plovers. Resident, but less common during the winter months. Occurs in pairs or family groups in open habitats, most often near water.
Length: 35 cm

WR PS NG

BETH PETERSON/AFRICAN IMAGES

Blacksmith Plover

Black, grey and white plover with dark red eyes. The long legs are grey-black. Occurs in pairs or family groups in open habitats, often near water. Like other plovers, they lay their eggs on bare ground and rely on egg and nestling camouflage for protection. When nests are threatened, the birds rise into the air above the intruders, chanting the metallic 'tink-tink' call which lends the bird its name.
Length: 30 cm

WR PS NG

Spotted Dikkop

Cryptically-coloured, plover-like bird with **long yellow legs** and very large **yellow eyes**. The back is streaked and blotched in fawn and brown, the underbelly is white. Nocturnal, but may be seen roosting under shrubs by day. A piercing, flute-like call is uttered after dark. The well-camouflaged eggs are laid on bare ground and they and the young are vigorously defended.
Length: 45 cm

NG PS

Greyheaded Gull

Grey and white gull with distinctive grey head only in the breeding season (late winter and early summer). The **red bill and legs** become brighter when breeding. The immature and non-breeding adult has a white head. Occurs in pairs or flocks at open water. Particularly common on the East Rand, including Johannesburg's international airport. Often scavenges from rubbish dumps.
Length: 42 cm

PS WR

Whitewinged Tern

Small, freshwater tern with distinctive breeding plumage of **black body, white upper wings and tail**, and red feet. In non-breeding dress it is pale grey on the back and wings with white underparts and head smudged in black behind the eye. Fairly common **summer migrant**. Flies above pans and dams, frequently dropping to catch small fish. Flocks rest on exposed shores.
Length: 23 cm ss: Whiskered Tern (25 cm)

WR

Swainson's Francolin

Stocky, dark-brown francolin with streaky plumage and **red facial mask**. Occurs in groups of three to five. Most active in early mornings or evenings. Sensitive to human disturbance, this ground-dwelling bird requires thick cover for refuge. The harsh, crowing call is made at dawn and dusk from an exposed position on a fence pole or rock.
Length: 38 cm ss: Coqui Francolin;
Orange River Francolin

NG S

Helmeted Guineafowl

Distinctive, charcoal-black bird spotted with white. The naked face is blue and red, and the elongated 'helmet' and hooked bill are horn-coloured. Occurs in flocks which are constantly on the move. The diet includes termites, ants and other insects as well as seeds and corms, but it is less inclined to dig than are most francolins. Flocks roost in trees at night or, occasionally, on overhead wires.
Length: 56 cm

PS NG S

Blackshouldered Kite

Small, mostly white raptor with pale grey back and wings; the black 'shoulders' are conspicuous in flight or at rest. The **eyes are bright red** and **feet yellow**. Immature birds are mottled grey-brown on the back. Occurs singly in open areas, perching on poles and telephone wires, and is commonly seen hovering with rapid wing beats before dropping onto prey. A stick nest is built in a tree.
Length: 33 cm

Yellowbilled Kite

Large, rusty-brown raptor with a **fork-shaped tail** which is often fanned out into a triangle in flight. The numbers of this **summer migrant** fluctuate from year to year. Roosts of several hundred birds are known in the Honeydew area. Common in the grounds of the Johannesburg Zoo, where they pirate food scraps from animal pens. The diet includes rodents, insects and carrion.
Length: 55 cm ss: Black Kite; African Marsh Harrier

Steppe Buzzard

Large, heavy-bodied raptor mottled in brown and fawn, with a **smudgy pale bar on the breast**. The cere and legs are yellow. Much individual variation in plumage occurs. A **summer migrant** from Europe and Asia, but much less common than the previous species in the Johannesburg area. Often perches on telephone poles along roadsides. Feeds on rodents and reptiles.
Length: 50 cm ss: Jackal Buzzard

Ovambo Sparrowhawk

Slender, dark-grey raptor with **white underparts finely barred in grey**. The legs and cere are yellow to pink. This secretive bird has adapted well to the alteration of the landscape, with copses of Grey Poplar *Populus canescens* providing ideal nesting sites and places in which to retreat. Birds – from warblers to doves – are caught after a chase.
Length: 35 cm ss: Little Banded Goshawk; Lizard Buzzard; Little Sparrowhawk

Greater Kestrel

Small, tawny-ginger raptor similar in size and habits to the Blackshouldered Kite. The body is **boldly barred in black**, and when seen from close range the **pale eye** is diagnostic. Often hovers above grassland before dropping onto prey such as grasshoppers, lizards and small rodents. The underwings are white. It breeds in old crow nests, often in tall electricity pylons.
Length: 35 cm ss: Rock Kestrel; Lesser Kestrel

Lanner Falcon

Medium-sized, silvery-grey raptor with buffy underparts. The **rust-red crown** and **dark hood extending as tear-marks** below the eyes are diagnostic. Immature birds are streaked on the chest. Flies extremely fast on pointed wings, catching birds up to its own size. Breeds in winter, in nests made by crows or other birds, often on pylons but sometimes even on the bare ledges of tall buildings.
Length: 40 cm ss: European Hobby; Peregrine

PS   KC   NG

Black Eagle

Very large, jet-black raptor, with a distinctive **white cross on its back**. The legs and cere are yellow. Immatures are mottled in brown and fawn. A pair breed on a cliff alongside the Witpoortjie Falls in the Witwatersrand Botanical Garden where they are strictly protected and monitored. Stragglers are known in the Klipriviersberg and elsewhere. Feeds primarily on Rock Dassies.
Length: 75 to 85 cm (females larger)

KC

Pied Crow

Large, shiny-black crow with distinctive **white breast and collar**. The legs, eyes and bill are black. As a scavenger, it has adapted well to man's surroundings where it feeds on road kills and at rubbish dumps. A stick nest is built in *Eucalyptus* trees or on pylons in early summer. May gather in flocks during winter. Speaks well in captivity but it is illegal to keep them as pets.
Length: 50 cm

KC   NG   PS

41

Whiterumped Swift

Black swift with a **deeply-forked tail** often held closed in flight. The **white rump** is conspicuous from above. In common with all swifts, it spends most of its time on the wing, hawking for tiny insects, and is physically unable to perch on wires or branches. Compared to swallows, the **wings are long and pointed to create a sickle-shaped outline**. Nests on buildings or cliffs. Absent in midwinter.
Length: 15 cm

Little Swift

Black swift with a **short square tail**, and large area of white on the rump. Less prone to fly as close to the ground as the Whiterumped Swift. Very gregarious, often gathering in large flocks to forage and breed. Water towers and concrete bridges are favoured nesting sites, but they are also partial to tall buildings in town and city centres. Present all year round.
Length: 14 cm

Palm Swift

Slender, **ashy-brown** swift with very long wings and pointed tail. There is no white on the rump. As its name suggests, it is associated with palm trees in which it makes its simple nest. Occurs in small flocks, often near water. All palm trees in the Johannesburg area have been planted by man, thus allowing this bird to expand its range into the region. Present all year round.
Length: 15 cm

Rock Martin

Plain-brown relative of the swallows with **broad wings and square tail**. Similar in colour to the Palm Swift but more stoutly built and flies more slowly. Small white spots are visible on the outspread tail. Occurs in the vicinity of rocky outcrops and cliffs, where it builds its nest under overhangs. Also breeds on buildings and under verandas. Present all year round.
Length: 15 cm

European Swallow

Glossy, royal-blue swallow with white belly and underwings. The **throat and forehead are russet**. Immature birds are dusky white and mottled on the throat. Present between late September and April, this is an abundant and familiar **non-breeding migrant** from Europe. Gathers in great numbers to perch on telephone wires each evening at the end of summer. Roosts in reedbeds.

Length: 18 cm ss: South African Cliff Swallow

PS WR S

Whitethroated Swallow

Glossy, royal-blue swallow with white underparts divided by a narrow **blue breast band**. The small russet forehead patch is obvious only from close range. An elegant, slow-flying swallow which occurs in pairs or small family groups. Present only in **summer**, it usually builds its nest under low bridges or culverts. Often seen near open water which it skims for insect prey.

Length: 17 cm ss: House Martin

MANFRED REICHARDT

PS WR

Greater Striped Swallow

Glossy, royal-blue swallow with off-white underparts lined with **smudgy black streaks**. The **crown and forehead are dull orange**. Easily confused with the next species but larger, and a more common bird in the Johannesburg area. A **summer migrant**, it nests under bridges, culverts and eaves in a closed, mud-pellet structure with a tubular entrance. Often feeds with other swallows.

Length: 20 cm

NG PS KC

Lesser Striped Swallow

Glossy, royal-blue swallow with off-white underparts lined with **bold black streaks**. The **crown, forehead and sides of face are bright orange**. Slightly smaller and more slender than the previous species, this swallow is a less common **summer migrant** to the area. Nests are identical to those of its larger relative. Often feeds with other swallows.

Length: 16 cm

PS S

43

Laughing Dove

Small, cinnamon-grey dove with a pale pinkish head and grey wings. A variable amount of black speckling forms a 'necklace' on the chest. One of the most common and confiding birds in the region. Occurs in pairs and is particularly abundant in parks and gardens. The nest is a fragile structure of twigs built on a branch or man-made structure.
Length: 25 cm

LEX HES

G PS NG S

Cape Turtle Dove

Pale grey dove with a brownish back. The distinctive **black neck collar is edged in white**, and the **eyes are black**. The **white-tipped tail** is obvious in flight. Occurs in pairs or flocks. The lovely 'kuk-cooor' call, repeated over and over again from a tree-top, is one of Africa's characteristic sounds. May be confused with the next species, but is considerably smaller and prefers more open habitats.
Length: 28 cm

NG S

Redeyed Dove

Pink-grey dove with a brownish back and **very pale head**. The neck collar is black, and the **red eyes** surrounded by bare red skin. The **tail is brown-tipped** above. Occurs in pairs. The mixed repertoire of calls includes a soft 'coo-coo-kuk'. May be confused with the previous species, but is considerably larger and prefers wooded habitats.
Length: 35 cm

G RF

Rameron Pigeon

Large, dark maroon pigeon with white speckles and spots on the wings and breast. The **bill, bare skin around the eyes, and feet are bright sulphur-yellow**. Occurs in pairs or flocks in well-wooded habitats. Feeds mostly on fruit, and is particularly fond of the alien Bugweed *Solanum mauritianum*. Often roosts on open branches of large trees. This is a recent immigrant to the region.
Length: 42 cm

RF G PS

Feral Pigeon *

Large, variably-coloured pigeon which is predominantly charcoal-grey. Many birds have pale grey backs with dark bands on the wings. Very gregarious, it is most common in the centre of Johannesburg where it feeds in the streets, gutters and small parks. Nests are built on the ledges of tall buildings. Descendent of the domesticated Rock Dove of Europe.

Length: 33 cm

PS

Rock Pigeon

Large, reddish-grey pigeon with bold **white spots on the wings**. The underparts and head are pale grey, and the **eyes are surrounded by bare red skin**. Occurs in pairs when breeding, but flocks may congregate in agricultural lands to feed on fallen grain. Usually nests on rocky ledges, but may also breed in the eaves of houses or in the mass of dry leaves on palm trees.

Length: 33 cm

PS   **KC**

Redchested Cuckoo

Pale-grey bird with long pointed wings. The **chest is rusty-red** and the underparts buffy with grey bars. Young birds are darker. Elusive **summer migrant**. The repetitive 'piet-my-vrou' call of the male is a familiar sound in midsummer, sometimes carrying on into the night. The female's call is a shrill 'pipi-pipipi'. Lays its eggs in the nests of other birds, particularly the Cape Robin.

Length: 30 cm

PS   **G**   **RF**

Diederik Cuckoo

Small vociferous bird with snow-white underparts. The male is **metallic green** above with **red eyes** surrounded by red skin. The female is bronze above, spotted in white. The immature bird has an orange bill. A **summer migrant** from central Africa. The name is derived from the repetitive call – 'di-di-deederik'. Eggs are laid in the nests of sparrows, weavers and bishops.

Length: 18 cm

RF   **WR**   **G**

45

Barn Owl

Pale, ghostly-looking owl with **white, heart-shaped face** and underparts. The back and wings are reddish-grey. Nocturnal and seldom seen. Rodents and shrews are the favoured prey, but small birds are also taken. Usually nests in old buildings, holes in cliffs and mine dumps. In the Johannesburg area it often breeds in the dry leaves of palm trees. The call is an eerie screech.

Length: 32 cm ss: Grass Owl

Spotted Eagle Owl

Large, grey-brown owl with dark blotches on the back and fine barring on the breast. The **ear tufts** are distinctive and the **eyes are bright yellow**. Nocturnal, but frequently seen in suburbs where it hunts under lights. Sits on roads, often with dire consequences. Breeds in holes in trees or among rocks. Feeds on a variety of small creatures, including winged termites. The call is a low, deep 'whoooo'.

Length: 45 cm

Marsh Owl

Dark brown owl with pale buffy under-parts and face. The eyes are black and the tiny ear-tufts barely noticeable. Nocturnal, but often active in the late afternoon or early morning. Sometimes perches on fence poles. Favours undis-turbed grasslands, often near wetlands. If disturbed, it flies in circles above the intruder. Nests on the ground among long grass. The call is a harsh croak.

Length: 35 cm ss: Grass Owl

Burchell's Coucal

Large, chestnut-red bird with black head and creamy-white underparts. Flies on broad wings in a floppy, unbalanced manner. In early mornings and evenings, it utters its bubbling call which sounds like liquid being poured from a bottle. Sometimes ventures onto lawns. Feeds on large insects, rodents and nestling birds. Related to the cuckoos, but it rears its own young.

Length: 44 cm

46

Grey Lourie

Large, pale grey bird with a **long lacy crest**. It often draws attention to itself with its harsh 'go-away' call. A fairly recent immigrant to the area, it is now common in many gardens and parks. Occurs in pairs or small flocks. Feeds on fruit. Partial to the berries of the alien Syringa *Melia azederach*, aiding its spread in the process. The small stick nest is often built in a conifer tree.
Length: 48 cm

G   PS   S

Redfaced Mousebird

Small, pale grey, mouse-like bird with a very long tail. The underparts are buffy. The **bright red facial skin** is diagnostic. Occurs in small flocks which **usually fly together**, and speedily, from one place to the next. Birds huddle together when roosting. Soft fruit and berries are the main food. The small stick nest is usually built within a tangled bush or creeper. Call is a clear whistle.
Length: 34 cm ss: Whitebacked Mousebird

BRENDAN RYAN

G   S

Speckled Mousebird

Small, dark grey, mouse-like bird with a very long tail. The underbelly is buffy. The face is black and the **lower bill white**. Unlike the Redfaced Mousebird, flocks **usually fly one-by-one**, in a panicky fashion from bush to bush. Birds huddle together when roosting. Favours tangled growth in which to roost and breed. Berries and soft fruit make up the bulk of their diet. Call is a harsh chatter.
Length: 35 cm ss: Whitebacked Mousebird

S   RF   G

Hoopoe

Distinctive, brick-red bird with bold black and white wings and **fan-shaped crest**. The long curved bill is used for probing the ground for worms and insects. In its slow lazy flight it resembles a giant butterfly. Young are reared in a hole in a tree or wall or a nest box. Pairs or family groups are often seen on lawns of larger gardens and parks. The call is a distinctive 'hoop-hoop'.
Length: 28 cm

PS   S   G

BRENDAN RYAN

Redbilled Woodhoopoe

Long-tailed, inky-blue bird with a metallic green sheen. The **long curved bill and short legs are coral red**. An active and noisy bird which lives in flocks of five or more. The cackling call is made in unison by flock members and often culminates with all the birds rocking back and forth on branches. Insect larvae are extracted from under the bark of trees. Young are reared in a tree hole.
Length: 36 cm

RF　G　PS

European Bee-eater

Multi-coloured bird with chestnut-gold back, turquoise-blue wings, sky-blue breast and yellow throat. The pointed wings are distinctive. A non-breeding **summer migrant**, this acrobatic bird occurs in flocks of up to 100. Flying insects, especially bees, are the favoured prey. Often hawks insects near water. The call is a frog-like 'pruuup'. Flocks usually roost in *Eucalyptus* trees at night.
Length: 30 cm ss: Whitefronted Bee-eater

NG　PS

JOHN CARLYON

Redthroated Wryneck

Small, cryptically-plumaged bird with **rusty-red throat** and thin black streak running down the centre of the head and back. Well camouflaged against tree bark. Occurs singly or in pairs. It feeds mostly on the ground, picking up ants and termites with its long sticky tongue. Nests in natural tree holes, nest boxes, or fence poles. The squeaky call is often made from the highest available perch.
Length: 20 cm

S　PS

Cardinal Woodpecker

Small woodpecker with an olive-green back with horizontal barring, and off-white breast with vertical streaks. Only the male sports the crimson crown. Occurs in pairs which move among branches, keeping contact with their 'kree-kree' call, and regularly drumming branches. Insect larvae are extracted with its long tongue. Nests in custom-made holes in hollow branches.
Length: 15 cm ss: Goldentailed Woodpecker

S　RF　PS

Blackcollared Barbet

Stocky barbet with a **crimson-red face** bordered by a **black collar**. The back is olive-green and the underparts buffy-yellow. The stout **black bill** is used to excavate nests in trees and nesting-logs. Noisy and conspicuous, it occurs in pairs or family groups. The call is a repetitive 'duduloo-duduloo' in duet or chorus. Insects, berries and fruit are eaten, and it regularly feeds at garden bird tables.
Length: 20 cm

RF PS G

Crested Barbet

Multi-coloured barbet with **yellow and red underparts and face**, and black back blotched in white. The ragged crest is raised in alarm. The stout **pale bill** is used to excavate nests in trees and nesting-logs. Often clashes with Indian Myna over nest sites. Occurs in pairs. The call is an alarm-clock trill. Berries, fruit and insects are eaten, and it regularly visits bird tables.
Length: 23 cm

S PS G

Acacia Pied Barbet

Small, black and white barbet with a diagnostic **scarlet forehead**. It is rather shy and rarely enters gardens. Fairly common in the acacia savanna of the Klipriviersberg. Hole nests are excavated in hollow branches and the stems of the Mountain Aloe *Aloe marlothii*. Fruit and berries, including those of mistletoe, are favoured. The call is a nasal 'nehh-nehh' or a hoopoe-like 'hoop-hoop'.
Length: 18 cm ss: Yellowfronted Tinker Barbet (11 cm)

S

Lesser Honeyguide

Small, olive-brown bird with distinctive **white outer tail feathers**. Lacking any bold colouring and secretive in behaviour, this bird is rarely noticed. In common with other members of its family, this bird is a brood parasite, laying its eggs in the hole nests of Crested and Blackcollared Barbets, and is regularly chased by these birds. Feeds on insect larvae and beeswax.
Length 15 cm ss: Sharpbilled Honeyguide; Greyheaded Sparrow

G RF S

Brownhooded Kingfisher

Medium-sized kingfisher with long, dusky-red bill and **turquoise-blue wings**. The back of the male is black, that of the female brown. Occurs singly or in pairs **away from water**, preying on large insects and lizards. Perches conspicuously and calls with a sharp, descending whistle, or a harsh 'klee-klee-klee' alarm. Eggs are laid in a self-excavated hole in an earth bank.

RF G

Length: 24 cm ss: Woodland Kingfisher

Pied Kingfisher

Medium-sized, black and white kingfisher with a long black bill. The sexes differ, with the male having a double black bar on the breast, and the female a single broken bar. Occurs in pairs or family groups at water, where it regularly hovers above the surface before plunging in after small fish. The call is a shrill twitter. Eggs are laid in a self-excavated hole in an earth bank.

WR

Length: 28 cm

Giant Kingfisher

Very large kingfisher with **charcoal back flecked in white**. The sexes differ, with the male having only the breast rufous, while the female is rufous on underbelly and underwings. Occurs in pairs or family groups, most often at moving water. Crabs are the favoured prey. The loud 'khak-khak-khak' call is often made while in flight. Eggs are laid in a self-excavated hole in an earth bank.

WR

Length: 46 cm

LEX HES

Malachite Kingfisher

Tiny kingfisher with **bright-blue back** and orange-fawn underparts. The long **bill and small feet are scarlet-red**. The crest is malachite-green flecked with black. Sexes are alike but the immatures are duller and have a black, not red, bill. Occurs singly or in pairs, keeping to reeds and sedges at the water's edge. Small fish, frogs, tadpoles and aquatic insects are caught. The call is a sharp whistle.

WR

Length: 14 cm

WILDERNESS SAFARIS/COLIN BELL

Rufousnaped Lark

Small, reddish-brown bird with a loose crest. Other than a pale eyebrow, it lacks any obvious plumage features. Its habit of singing from a conspicuous fence pole or low bush is helpful in identification. The call is a drawn-out 'tseeu-tseeuoo' whistle. Forages for insects on the ground. Restricted to open grassland, this species is most often seen in the northern and western parts of the area.

Length: 18 cm ss: Clapper and Longbilled Larks

WILDERNESS SAFARIS/COLIN BELL

NG

Grassveld (Richard's) Pipit

Small brown bird with pale fawn underparts and distinctive **white outer tail feathers** most obvious in flight. The breast is finely streaked. Much slimmer than the previous species. Occurs in open areas of short grassland as well as in the vicinity of pans and even on sports fields. The call is a 'chri-chri-chri' whistle, uttered in an aerial display flight. Mostly silent in winter.

Length: 16 cm ss: Longbilled Pipit; Buffy Pipit

PS WR NG

Cape Wagtail

Small, grey-brown bird with long tail. The underparts are off-white and there is a thin bar on the throat. True to its name, the tail is constantly bobbed up and down. Confiding and adaptable, it is often more common around human settlements than in its traditional waterside habits. Also lives in streets and carparks of town centres. Feeds on small insects and vulnerable to poisoning by pesticides.

Length: 19 cm ss: Yellow Wagtail

WR PS G

Orangethroated Longclaw

Small, grey-brown bird with upright stance. Seen from behind, it appears a drab, unremarkable bird, but the belly and breast are sulphur-yellow and the **throat brilliant orange**. The reason for the exceptionally long claw on the hind toe is unknown. The call is a cat-like 'meauw'. Occurs in pairs or family groups in undisturbed grasslands. Now restricted to the outskirts of suburbs.

Length: 20 cm

MANFRED REICHARDT

NG

Blackeyed Bulbul

Small brown bird with black head, pale underside and **lemon-yellow vent**. Occurs in pairs or family groups in a variety of habitats. A common and confiding resident of many gardens. The call is a series of liquid whistles, or a harsh 'tshwit-tshwit' alarm made when cats, owls or snakes are about, or when going to roost in the evening. The diet includes berries, fruit and insects.
Length: 22 cm

G RF PS

Cape Robin

Small, grey-brown robin with **orange tail and throat**, and **white eyebrow**. The sexes are alike, but the immature is buffy and speckled. Occurs in pairs, spending most time on the ground, but retreats quickly to cover if disturbed. May become confiding in gardens. Worms and insects are captured on lawns and among leaf litter. The call is a melodious whistled song or a throaty alarm.
Length: 18 cm

RF G

Olive Thrush

Small, olive-brown thrush with distinctive, **orange-yellow bill and legs**. The underbelly has a variable amount of dull orange. The sexes are alike, but the immature is paler and speckled below. Occurs in pairs and is most active at dawn and dusk. Forages on the ground for worms and insects, but also feeds on berries and nestling birds. The call is a flute-like whistle.
Length: 24 cm

RF G PS

Cape Rock Thrush

Small, reddish-brown thrush restricted to rocky habitats. The male has **orange-ochre underparts** and a **blue-grey head**. The female is paler below with a dusky-brown head. Occurs in pairs, often perching in an upright position on an exposed rock. The call is a clear flutey whistle. May forage and nest around houses where these are built near ridges or koppies.
Length: 21 cm

52

Mountain Chat

Variably-plumaged chat of rocky habitats. The male may be predominantly black or pale grey, but always has a distinctive white shoulder patch and a **white V-pattern on the rump and tail**. Some birds have a white cap. The female is brown with a similarly patterned rump and tail. Frequently sits on rooftops of houses near rocky areas. Insects make up the diet. The song is a lively whistle.
Length: 20 cm

BRENDAN RYAN

KC

Mocking Chat

Dark chat of rocky habitats. The male is blue-black on the head, back and wings, with a **bright chestnut rump and belly**. The white shoulder patch is variable in size. The female is charcoal-grey on the back with rusty underparts. As its name suggests, it is an accomplished mimic of other birds' songs. Less common than the Mountain Chat but very inquisitive and confiding.
Length: 23 cm

KC

Familiar Chat

Small, pale brown chat with a **rusty-red tail** most obvious in flight. Its habit of flicking its wings on alighting is an aid to identification. The sexes are alike. Occurs in pairs in open habitats and farmlands north and south of the suburbs. Like all chats, it is insectivorous, feeding on small insects and caterpillars taken on the ground. It is often very confiding, and may nest near houses.
Length: 15 cm

KC NG

Stone Chat

Small, round-bodied chat. The male is brown on the back with a black head, white collar and rufous breast. The female is paler with a fawn breast. Fences and low bushes are favoured perching places, making this a conspicuous bird wherever it occurs. Occurs in pairs in open grassland and road verges, and often in rank growth near water. Small insects are caught on the ground.
Length: 14 cm ss: Capped Wheatear

WR

NG

53

Blackchested Prinia

Tiny, pale brown bird with off-white eyebrow and long tail often held upright. Breeding adults have off-white underparts with a **black band on the chest**. Non-breeding adults lack the breast band and are lemon-yellow below. Occurs in pairs or family groups, usually away from water. Inquisitive and noisy, the call is a loud, repetitive 'zit-zit-zit', uttered from a grass stem or bush.
Length: 14 cm

Tawnyflanked Prinia

Tiny, pale brown bird with off-white eyebrow and long tail often held upright. Adults are off-white below with tawny flanks. The **rusty-red wings** are diagnostic. Occurs in pairs or family groups in rank or marshy growth along streams, or on the fringes of wetlands. Inquisitive and noisy, the call is a loud, repetitive 'teee-teee-teee' or 'zinc-zinc-zinc' uttered from a grass stem or bush.
Length: 14 cm

Willow Warbler

Tiny, pale brown warbler with **off-white eyebrow**. Some individuals are pale olive above with yellowish underparts. Unobtrusive and mostly silent non-breeding **summer migrant** from northern Europe. Forages alone or with other birds in leafy vegetation, feeding on small insects. Active, but quiet and easily overlooked. Unlike many similar warblers, it is not found in reedbeds.
Length: 11 cm ss: Garden Warbler (14 cm)

Cape Reed Warbler

Sizeable, pale brown warbler with creamy-white underparts. The **pale eye-stripe** is distinctive. Occurs singly or in pairs in **reedbeds and rank growth** around permanent water. Less shy than other similar warblers which share this habitat. The flutey, robin-like call – 'chirrup-chee-treee' – is more musical than that of other wetland warblers.
Length: 17 cm ss: African Marsh Warbler (13 cm); European Reed Warbler (13 cm); Great Reed Warbler (19 cm); African Sedge Warbler (17 cm)

Neddicky

Tiny, plain brown cisticola with a **russet crown** and short tail. There is no distinct eyebrow. Occurs in pairs or family groups on **rocky hillsides**, often perching conspicuously on low bushes or boulders. Inquisitive, it sometimes approaches or follows people who enter its territory. Insects such as termites and ants are the main food. The call is a persistent, squeaky 'tsee-tsee-tsee-tsee'.
Length: 11 cm

MANFRED REICHARDT

S   KC

Fantailed Cisticola

Tiny, sandy-brown cisticola with dark streaking on the back and crown. The tail is short, and often fanned out. Occurs in pairs or groups in **open grassland, often near water**. Similar to the two 'cloud-scraper' cisticolas listed below, but distinguished from them by its 'zit-zit-zit' call and dipping aerial display. The 'cloud-scrapers' display at a greater height and snap their wings.
Length: 11 cm ss: Cloud Cisticola; Ayres' Cisticola

NG   WR

Levaillant's Cisticola

Small, reddish-brown cisticola with **black back feathers** edged in brown. The long tail is reddish and tipped in off-white. The cap is red-brown. Occurs in pairs or small groups in **rank growth near permanent water**, but avoids dense reedbeds. More boldly marked on the back than the Wailing Cisticola – the two birds do not share the same habitat. The call is an insistent 'dswe-dswe-dswe'.
Length: 14 cm

WR

Wailing Cisticola

Small, fawn-brown cisticola with **grey back feathers** edged in brown. The long tail is brown and tipped in buff. The cap is red-brown. Occurs in pairs or small groups on **grassy hillsides with exposed rocks** – it avoids rank growth near water. Less boldly marked on the back than Levaillant's Cisticola. Inquisitive and noisy. The call is a plaintive 'whee-whee-whee'.
Length: 13 cm ss: Lazy Cisticola (plain back)

S   KC

55

Fiscal Flycatcher

Small, dark-backed flycatcher with white underparts. The male is jet-black above, the female dusky-brown. Both sexes have **white wing patches** and the **tail is edged in white**. Occurs in pairs, and can be confiding. Often confused with the longer-tailed Fiscal Shrike (opposite), which it is said to mimic. The call is a soft but pleasing song. Feeds on flying insects.
Length: 19 cm

  PS

Spotted Flycatcher

Small, pale brown flycatcher with white throat and breast finely streaked in brown. The name is misleading, as the adult is not spotted at all (the immature is, but so are the young of many other flycatchers). The **faint streaks on the forehead** are diagnostic. It invariably **flicks its wings on alighting**. Perches on a low branch or fence pole, launching out to catch insects. Non-breeding **summer migrant**.
Length: 14 cm

Paradise Flycatcher

Small flycatcher with **rusty-red back** and grey-blue underparts. The bill and eye-wattles are blue. The male has a very long, **ribbon-like tail** giving him the appearance of a small rocket when in flight. The female has a much shorter tail. The call is a musical warble or a 'chweet-chweet' alarm. The tiny 'egg-cup' nest is built on a bare branch, often over water. **Summer migrant**.
Length: 22 cm (plus 18 cm tail in breeding male)

RF G

Fairy Flycatcher

Tiny flycatcher with pale grey upper-parts and **black mask topped with a white eyebrow**. Warbler-like in habits, it gleans small insects from the leaves rather than capturing them in an aerial pursuit. On its breeding grounds in the south-western Cape and Karoo it occurs in pairs or family groups, but it is solitary on the Highveld. Quiet and easily over-looked. **Winter migrant**.
Length: 12 cm ss: Chinspot Batis

Fiscal Shrike

Black and white shrike with **long tail** and **white V-pattern on the back**. Females have a rust wash on their flanks. Young are mottled and barred in brown and white. Perches conspicuously, in **upright posture**, on fences, wires, walls and outer branches of trees. Feeds mostly on large insects, spearing victims onto thorns or barbed wire – to be retrieved later.
Length: 23 cm

NG PS G

Southern Boubou

Black and white shrike with **buffy underparts**. With a **white V-pattern on the back** it is superficially similar to the Fiscal Shrike but the tail is shorter. Sexes are alike. The body is held in a horizontal, rather than upright, posture. Occurs in pairs which keep mostly to the interior of low shrubs and dense foliage. The calls are varied, including liquid or grating notes usually uttered in duet.
Length: 23 cm

G RF

Puffback

Small, black and white shrike with white edges to wing feathers. The female is duller with white cheeks and eyebrows. Both sexes have **red eyes**. The male puffs out a mass of white feathers on his rump during courtship – hence the name. The call is a loud 'chick-wheeo'. Keeps to dense foliage, often feeding alongside other small birds. Small insects are the main food.
Length: 18 cm ss: Brubru

G RF

Bokmakierie

Large shrike with olive-green back, grey head and **sulphur-yellow underparts**. The **black gorget** is diagnostic. Sexes are alike. Immature is duller with no gorget. Occurs in pairs in open habitats including rocky hillsides. The repertoire of calls is varied. A loud, ringing duet, beginning 'bok-bok-cheet', probably gives the bird its name. Feeds on beetles and other large insects.
Length: 23 cm

KC G S

57

JOHN CARLYON

Indian Myna *

Chestnut and black bird with white wing 'windows' and **mask of bare yellow skin**. Occurs in pairs or 'gangs' in the vicinity of human habitation or industry. This **alien species** spread to Johannesburg in the early 1950s. Disliked for its habit of nesting in eaves and of displacing indigenous birds. Eradication schemes have been unsuccessful. Feeds on scraps, insects and fruit. Call is a rasping hiss.
Length: 25 cm

PS **G**

Pied Starling

Olive-brown starling with **yellow gape** at the base of the bill, **off-white vent** and **pale eyes**. Occurs in groups of a dozen or more, but may form larger flocks in winter. It may clash over food or nesting sites with the previous species. On farm plots, it frequently feeds among domestic stock, snatching up disturbed insects. Breeds in holes in earth banks, less often in eaves. Call is a loud screech.
Length: 27 cm

NG **PS** **G**

Redwinged Starling

Blue-black starling with distinctive, **rust-red wings most visible in flight**, and **dark eyes**. The female has an ashy-grey head and is streaked on the chest. Occurs in pairs or small flocks in rocky habitats but has adapted to breeding on buildings and under bridges. Would probably be more common if the Indian Myna was not present. Feeds on insects and berries. Call is a liquid, flutey whistle.
Length: 26 cm

Glossy Starling

Metallic **blue-green** starling with bright **yellow eyes**. The sexes are alike. Immature birds are duller. Occurs in pairs when breeding and small flocks during winter. Probably impacted on to a greater extent than any other by the aggressive Indian Myna which shows a preference for similar breeding holes. Insects, berries and soft fruit are the main food. Call is a squeaky warble.
Length: 25 cm

S **PS** **G**

Cape White-eye

Tiny, pale green bird with a distinctive **ring of small white feathers around the eyes**. The sexes are alike. Occurs in pairs during the summer breeding season but flocks form up in winter. Feeds on small insects, such as aphids, as well as berries. Often visits gardens; fond of bathing in bird-baths or under sprinklers. The call is a soft, musical warble. Flocks roost in a huddle at night.
Length: 12 cm

LEX HES
(S) (RF) (G)

Black Sunbird

Medium-sized sunbird with sexes not alike. The male appears completely black, but closer investigation reveals the **amethyst throat and rump**, and **emerald forehead**. The female is brown on the back with pale grey underparts. The nest is a purse of leaves bound with spider webs. Feeds primarily on the nectar of tubular flowers. The call is a series of twittering notes.
Length: 15 cm

BRENDAN RYAN
(G) (RF)

Whitebellied Sunbird

Tiny sunbird with sexes not alike. The male is metallic green on the back and head with a purple-blue throat and **snow-white belly**. The female is grey above, with unstreaked white underparts. The nest is a purse of leaves bound with spider webs. Feeds primarily on the nectar of tubular flowers, especially aloes. Often enters gardens. The call is a strident tinkling warble.
Length: 11 cm ss: Lesser Doublecollared Sunbird

LEX HES
(G) (S)

Malachite Sunbird

Large sunbird with sexes not alike. The breeding male is emerald-green with a long tail and yellow 'epaulettes'; when not breeding, he becomes olive-brown with some green patches. Females are pale olive with yellowish, lightly-streaked underparts. Occurs in pairs or groups in **protea savanna**. May enter gardens to feed from nectar-rich plants and bathe under garden sprinklers.
Length: 25 cm (male) 15 cm (female)

MANFRED REICHARDT
(G) (S)

LEX HES

Cape Weaver

Sizeable weaver with **long pointed bill**. The sexes are not alike. Breeding male is golden-yellow with **ginger forehead and throat**. Females are pale olive on the back and creamy-yellow below. Both sexes have **pale eyes**. Insects and seeds are eaten. Woven nests are built among reeds or suspended from branches (often Weeping Willows) over water. Call is a swizzling buzz.

Length: 17 cm

Masked Weaver

Small weaver with **short pointed bill**. The sexes are not alike. Breeding male is sulphur-yellow with a **black mask** and **red eye**. The female is pale olive on the back and creamy-yellow below. Seeds make up the bulk of the diet. Woven nests are usually hung over water. Their habit of stripping the leaves of palms for nest-material has earned them the wrath of gardeners. Call is a swizzling buzz.

Length: 15 cm

Golden Bishop

Small finch with sexes not alike. The breeding male is predominantly **black with a brilliant yellow crown and rump**. In flight, he resembles a large bumblebee. Females are streaked in fawn and brown above with paler underparts and pale eyebrows. In winter, males resemble the drab females. Occurs in small groups in summer, but forms large flocks in winter. Call is a rasping buzz.

Length: 12 cm

MANFRED REICHARDT

Red Bishop

Small finch with sexes not alike. The breeding male is **crimson-red with a black face and underbelly**. Females are streaked in fawn and brown with paler underparts and pale eyebrows. In winter, males resemble the drab females. Colonies consist of several hundred. Chattering sizzles resound in the colony as rival males defend nests which are hung between reed stems.

Length: 14 cm

Longtailed Widow

Medium-sized finch with sexes not alike. The breeding male is black with elaborate tail feathers and **orange shoulder patches**. Non-breeding male resemble the drab, fawn and brown female but retains the orange shoulder patch. Occurs in small flocks in reedbeds and rank growth alongside wetlands. May also gather along roadsides to feed on seeding grasses.

Length: 19 cm (plus 40 cm tail in summer)

NG WR

Redcollared Widow

Small finch with sexes not alike. The breeding male is completely black with very long tail feathers and a narrow **red throat collar**. The non-breeding male resembles the drab, fawn and brown female. Occurs in small groups with a single male and his harem of females. Favours rank growth near wetlands but may also occur in dry savanna.

Length: 15 cm (plus 25 cm tail in summer)

WR S NG

Pintailed Whydah

Tiny finch with sexes not alike. The breeding male is black and white with long tail feathers and a bright **pink bill**. Non-breeding males and females are drab fawn and brown with striped crown and pink bill. The male draws attention by hovering above his harem of females, and chasing off other birds. Eggs are deposited in the nests of the Common Waxbill which rears the nestlings.

Length: 12 cm (plus 22 cm tail in summer)

G S NG

Redheaded Finch

Small, pale brown finch with sexes not alike. The male has a **dull red head** and **scaly feathers on the underparts**. The female has a grey-brown head and is barred on the pale underparts. The bill is pale grey and very stout. Occurs in pairs or small flocks. A recent immigrant to the Johannesburg region where it is most common on the East Rand. Most frequent during the dry winter months.

Length: 13 cm ss: Redbilled Quelea

G NG

61

LEX HES

Cape Sparrow

Small, **chestnut-backed** sparrow with white underparts. The male has a black and white face. The female is paler over-all with a grey and white face. Occurs in pairs during the summer breeding sea-son. The untidy straw nests are built in bushes or on buildings. Large flocks may form in winter. Feeds on seeds and scraps. Very familiar bird in gardens, and also present in city parks and on streets.

Length: 15 cm ss: Greyheaded Sparrow

House Sparrow *

Small, grey-brown sparrow with a streaked back. The male has a **grey crown and rump**, white cheeks and black throat. The female has no distin-guishing features other than a thin pale eye-stripe. This **alien species** occurs in pairs or small groups around buildings and other man-made structures, and on the pavements and gutters of central Johannesburg.

Length: 14 cm ss: Greyheaded Sparrow

Rock Bunting

Small brown finch with **black head striped in white**. The male is darker and more boldly marked than the female, with a **cinnamon-rufous breast**. Occurs in pairs on wooded hill slopes. The call is a grating swizzle, made from an exposed perch. Diet includes seeds and insects. The related **Cape Bunting** is grey on the crown and back, and off-white below.

Length: 14 cm ss: Cape Bunting

LEX HES

Streakyheaded Canary

Small, grey-brown canary with off-white underparts. The broad **white eye-stripe** and **streaked crown** are diagnostic. The rump is not yellow, as is the case with most other canaries. The sexes are alike. Occurs in pairs or small flocks, most frequently in protea savanna. The call is a repetitive, gentle song, usually uttered from a tree top. Feeds on seeds and small insects.

Length: 15 cm

Blackthroated Canary

Tiny, grey-brown canary with distinctive **lemon-yellow rump** – only visible in flight. The male differs from the female in having a **black smudge on the throat**. Both sexes are heavily streaked on the back. Occurs in small flocks. Feeds mostly on seeds, but also termites. Like other seed-eaters, it drinks regularly and may visit garden bird baths. Call is a jumble of soft musical notes.
Length: 12 cm

MANFRED REICHARDT

S G NG

Bronze Mannikin

Tiny, chocolate-brown finch with white underparts, black head and **pale lower bill**. The iridescent, green-bronze shoulders are not always obvious. Sexes are alike. Immature birds are buffy-fawn. Occurs in pairs or small flocks. Very restless. Feeds on grass seeds, often picking these up from the ground, and will also visit bird tables. Drinks regularly. Call is a soft, rasping warble.
Length: 9 cm

LEX HES

G S WR

Common Waxbill

Tiny, pale grey finch with rosy under-parts and long tail. The bill and eye-mask is bright pink. The whole body is finely barred. Sexes are alike. Immature birds are paler, with a black bill. Occurs in small flocks in rank growth in the vicinity of water, and often in the company of Bronze Mannikin or the following species. Feeds on fresh grass seeds but may also take insects.
Length: 13 cm

LEX HES

G WR

Orangebreasted Waxbill

Tiny, grey-backed finch with yellow-orange underparts. The **bill, eye-stripe and rump are scarlet-red**. Male is much brighter in colour than the female and immature. Occurs in small flocks in rank vegetation fringing wetlands. Flies off when approached, but usually settles a short way off. Feeds on fresh seeds but may also take small insects.
Length: 10 cm ss: Quail Finch

WR

Reptiles

A surprising variety of snakes, lizards and other reptiles occur in the Johannesburg area. All are cold-blooded, and require food less often than mammals or birds. Many reptiles become dormant during cold weather and several species hibernate during winter. They are among the most misunderstood and feared of all animals, which is a pity as most are fascinating and harmless.

Most snakes are nocturnal and seldom seen. Several potentially lethal venomous species do occur in the area (and are marked with a 🐍 symbol) but they will usually bite only in defence, and then as a last resort. If confronted by a dangerous snake, the best strategy is to remain calm and allow it every possible avenue of escape; any attempt to catch or kill it will only increase your chances of being bitten.

With only a few exceptions, reptiles are difficult to study in the wild. They are usually encountered by chance, but certain species are restricted to particular habitats and may be actively looked for. However, the practice of turning over rocks may lead to close encounters with adders, or interrupt the breeding or hibernation cycles of certain species. The capture and keeping of reptiles is illegal – most eventually die when taken out of their natural habitat. One way to learn how to identify various snakes and other reptiles is to visit the Transvaal Snake Park or Pretoria Zoo where many species are on display.

The names used in this section follow those in the *Field Guide to the Snakes and other Reptiles of Southern Africa* by Bill Branch (Struik, 1988) – the most comprehensive and compact reference book currently available.

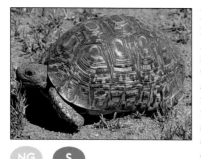

Leopard Tortoise

Largest and most widespread of South African tortoises. Adults weigh between 8 and 12 kg, but may reach 40 kg in captivity. It spends the day moving slowly about a home range of between 1 to 2 km², feeding on plant foliage and berries. The female lays a clutch of ping-pong ball-sized eggs in a shallow burrow. Young are vulnerable to many predators, and adults to grass fires and pet collectors.
Length: 30 to 45 cm (max. 72 cm)

LEX HES

Marsh Terrapin

Aquatic species, distinguished from the Leopard Tortoise by its flattened shell. Immatures have a white throat. Several individuals may occupy a suitable dam or stretch of stream, where they bask on rocks. May move over land, particularly after rain. Feeds on a wide variety of other animals including frogs, tadpoles and ducklings, and readily scavenges. Oval eggs are deposited in a burrow.
Length: 20 to 30 cm

Cape Dwarf Gecko

Tiny, pale brown gecko with a **thin tail**. **Active by day**. It lives on the trunks and branches of trees and also on the outdoor walls of houses. Like other geckos, the toes have tiny clinging scales which enable it to walk on vertical surfaces. Termites, ants and other small insects are the prey. Males are territorial. A pair of tiny, hard-shelled eggs are laid in cracks or crevices.

Length: 6 to 7 cm

G　S　KC

Transvaal Gecko

Small, mottled gecko with a **fat tail**. The very large eyes, set within the chubby head, indicate its **nocturnal** habits. Spends the day under stones and logs, emerging to hunt for moths, termites, beetles and other insects. Occasionally takes up residence in houses. The immature is more finely spotted. Much variation in colour occurs.

Length: 8 to 12 cm

WD HAACKE

KC　S

Flapnecked Chameleon

Distinctive reptile with large rounded head, **conical eyes** and **curled tail**. Adults are predominantly green but able to change colour to match their background. Slow-moving, they are **active by day**, often moving after rain. A clutch of 25 to 50 eggs is buried in the soil. Flies and other insects are caught with its elastic tongue. Is becoming scarcer due to pesticide use; many are killed on roads.

Length: 20 to 24 cm

KC　S　NG

Ground Agama

Plump, ground-dwelling lizard **active by day**. The large triangular head and protruding eyes are diagnostic. The male is conspicuous during the summer when it sits in exposed places, showing the **blue sides to the head**. In this position, it may bob up and down as though doing press-ups. Eggs are laid in sandy soil. Feeds on ants, termites and other small insects. Usually wary.

Length: 15 to 22 cm ss: Southern Rock Agama

G　NG　KC

Striped Skink

Slender, dark-bodied lizard with a long tapering tail. A **single pale stripe** runs on either side of the almost black body; the underbelly is yellowish. **Active by day**, it is found in a wide variety of habitats, and may be common on garden walls and paving around houses. Insect prey is captured after a chase. Young are born live rather than from eggs.

Length: 18 cm ss: Variable Skink (12 cm)

Cape (Threelined) Skink

Chubby, pale-bodied lizard with long tapering tail. **Three off-white stripes** run the length of the grey-buff body; the underbelly is off-white. **Active by day**, it occurs in open habitats where it hunts insects such as small beetles and termites. It burrows in loose sand but may also take refuge under rocks or on dried aloe stems. Interestingly, the young may be born live or out of eggs.

Length: 25 cm

Transvaal Girdled Lizard

Squat, scaly lizard with a triangular head and **thick spiny tail**. The upperparts are pale yellowish in colour, often with some striping and blotching. **Active by day**, but remains mostly within the cover of rocks, emerging to catch beetles and other insect prey. To escape enemies – including man – they wedge themselves into crevices, hooking the scales on their backs to the rock. One or two live young are born.

Length: 14 to 17 cm

Water Monitor (Leguaan)

Very large aquatic lizard with **long muscular tail** used in swimming or defence. Adults are dark olive-brown or grey-brown on the back, and paler below. Juveniles are boldly striped in yellow and black. Feeds on a wide variety of animals including crabs and frogs; partial to birds' eggs. Females lay 20 to 60 soft-shelled eggs in a hole in an active termite mound. Active by day.

Length: 1 to 2 m ss: Rock Monitor

Brown House Snake

Medium-sized snake with squared-off snout and **two pale lines running on either side of the pale eye**. Body colour is variable, ranging from light brown to rust. Prey is killed by constriction, then swallowed head first. **Nocturnal**. Up to 15 eggs are laid in summer. One of the few snakes which has adapted to the urban environment, it is harmless and useful because of its taste for rats and mice.

Length: up to 1 m ss: Aurora House Snake

C & T STUART

S G NG

Mole Snake

Large, thick-bodied snake with **short pointed snout**. The colour varies from ochre or grey to brick-red. Juveniles are attractively patterned in tan and black. **Active by day**, it preys on burrowing mammals in their holes. Victims are constricted before being swallowed. Non-poisonous, but very aggressive and able to inflict severe bites. Gives birth to 30 to 50 live young in autumn.

Length: up to 1.5 m (max. 2 m)

NG S

Common Egg-eater

Small, heavily blotched snake, said to mimic the poisonous Common Night Adder (p. 68). Frequently associated with termite mounds in which it shelters during the day. In a remarkable dietary adaptation, the jaws are able to be dislodged from the neck bones so that birds' eggs, larger than its own head, may be swallowed. It lacks sizeable teeth. Up to 25 eggs are laid.

Length: 50 to 70 cm (max. 1 m)

C & T STUART

WR S NG

Herald (Redlipped) Snake

Small slender snake with glossy, blue-black head and **orange-red upper lip**. The body is uniform olive-brown above, sometimes with small white spots; the underbelly is off-white. This snake has adapted well to suburbia and is some-times uncovered among garden litter or rubble. **Nocturnal**. It feeds mostly upon frogs so is most at home in wetlands. Up to 12 eggs are laid in summer.

Length: 60 to 75 cm (max. 80 cm)

NG G WR

Striped Skaapsteker

Small slender snake patterned with olive and black stripes and with a pure **white underbelly**. It feeds mostly on frogs, so is usually found in the vicinity of water. **Active by day**, it may be seen basking on roads before dashing for cover. Although poisonous to its small prey, it is incapable of injecting enough venom to kill anything larger, so the common name – 'sheep-stabber' is a misnomer.

Length: 70 cm ss: Spotted Skaapsteker

Rinkhals 💀

Large, **brown or black** snake with **rough scales**. The throat is marked with white bands. A relative of the cobras, this dangerous snake more often **sprays venom** than bites. It is slow-moving and has poor eyesight, a combination which often leads to contact with people. May sham death when threatened. Feeds mostly on toads, but will also take rodents. Up to 30 live young are born in late summer.

Length: up to 1.5 m

Common Night Adder

Small, stout snake geometrically-patterned in brown, tan and buff. The forward-pointing **V-pattern on the top of the head** – a trait shared by the Common Egg-eater – is obvious. It bites readily and is mildly poisonous. Nocturnal, it preys mostly on toads. Occurs in a variety of habitats. The only adder not to give birth to live young, up to 26 eggs are laid in summer.

Length: 40 to 60 cm (max. 1 m)

Puff Adder 💀

Large stout snake speckled in ochre, tan and black with conspicuous A-shaped chevrons. The scale tips are raised to give a rough appearance. Sluggish and well camouflaged this is the most dangerous snake in the area, and its bite may prove fatal. Although most active at dusk and after dark, it may be encountered during the day. Rodents, toads and ground birds are taken. Litters are of 20 to 40 young.

Length: 70 to 90 cm (max. 1.2 m)

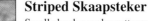

Frogs

Frequently encountered, but largely ignored, frogs are fascinating creatures deserving of closer study. They are exceptional in that they have two stages in life: the tadpole, which is usually totally aquatic, and the four-limbed adult, which may be active in or out of water. Many Highveld species have adapted their lifestyles to the general scarcity of suitable water bodies – they are dormant in winter, but emerge to breed at the onset of the summer rains. Frog populations are reduced by factors such as urban development and chemical pollution.

Like birds, frogs may be identified by their calls and this greatly facilitates their location and study. Most species are only active after dark, however, and are best found by going out with a strong torch after, or during, rain. Following up on calls can be rewarding, but requires great patience as frogs are able to detect vibrations made by moving humans. Gumboots are recommended when out 'frogging' and one should always keep an eye open for snakes – which have as much interest in frogs as any naturalist.

Keeping frogs in captivity can be most interesting. An aquarium can be used to observe tadpole development, and a glass terrarium for watching adults. Ensure that an emergent rock, or clump of plants, permits your captives to rest out of the water, and feed them with captured insects. When your observations are complete, release the frogs back into an appropriate habitat.

The names used in this section follow those in *South African Frogs* by Neville Passmore and Vincent Carruthers (Southern & Wits. Univ. Press, 1995). A tape entitled *Voices of South African Frogs* supplements this book.

Common Platanna

Extraordinary frog with a compressed body and **eyes on the top of its head**. Usually seen suspended just below the water surface. It is almost totally aquatic but may move on land during rain. The fore limbs are short and slender, but the hind legs are large and powerful with webbing between the **clawed toes**. Preys upon small fish, insects and tadpoles, and will also scavenge. Said to call underwater.
Length: 5 to 8 cm

VINCENT CARRUTHERS

WR

Giant Bullfrog

Very large, olive-green frog with **orange armpits** and **yellow throat**. Adults bite if handled. Remains underground during winter, but emerges after heavy summer rains to occupy temporary pans in grassland. Adults then fight, mate and lay large numbers of eggs which develop rapidly. Preys on other frogs and rodents. The call is a deep 'whooop'. Once common on the Highveld, but now rather rare.
Length: 8 to 20 cm

VINCENT CARRUTHERS

WR   NG

Guttural Toad

Sandy-brown or grey toad with lumpy skin. May be confused with the next species but distinguished by the **red flecks on the thighs**. Breeds in permanent water (including garden ponds) but often found some distance from water. The call is a reverberating 'snore' often uttered in chorus, and beginning in early spring. Strictly nocturnal. Breeding is stimulated by rain.

Length: 5 to 7 cm

Raucous Toad

Sandy-brown or grey toad with lumpy skin. Distinguished from the previous species by the **absence of red flecks on the thighs**. This species is more often found alongside running water than at ponds and dams, but may occasionally venture into dry rocky country. Strictly nocturnal. The call is a sharp 'kwaak', repeated at intervals. Does not call in large choruses.

Length: 5 to 7 cm

Red Toad

Grey or brick-red toad distinguished by a **raised dark line down its side** from eye to groin. Two dark spots may be present on the back. Breeds in large numbers in temporary pans and dams where males indulge in incessant and deafening bouts of chorusing. The call is a deep 'ooomp' made whilst afloat in water. Nocturnal, but may be active on cloudy days. Frequently found in gardens.

Length: 4 to 7 cm

Natal Sand Frog

Tiny grey or brown frog with a toad-like shape and posture. The dark patch behind the eye is not always present. It spends most of each year buried beneath the ground, emerging after rain to breed in shallow pools and puddles in grass-land. It may also occur along streams or in permanent wetlands. The call is a high-pitched note made in short choruses.

Length: 4 cm   ss: Tremolo Sand Frog

Common River Frog

Slender, green or brown frog with dark spots and blotches. The **snout is pointed** and the **hind legs are very long and powerful**. Occurs in perennial streams and permanent dams and ponds. Often sits on emergent stones or at the water's edge, but is quick to dive in if disturbed. The call is a distinctive rattle and croak, uttered even on cold winter nights.

Length: 5 to 7 cm ss: Cape River Frog (10 cm)

VINCENT CARRUTHERS

Striped Stream Frog

Very small slender frog with **pointed snout** and **extremely long toes**. The body is straw-coloured with bold, dark-brown stripes running down the back. Occurs in long grass and sedges on the fringes of streams, dams and other bodies of water. Secretive and difficult to find, it rarely enters water except to breed. The call is a sharp 'pip', uttered singly or in sequence.

Length: 4 cm

VINCENT CARRUTHERS

WR

Common Caco

Tiny frog with **extremely variable colouration**. Individuals may be shades of green or brown with, or without, stripes, blotches and spots. If caught, the **white underbelly with dark spots** is diagnostic. It is only active after rain. Numbers gather in flooded grassland to breed, calling with an explosive 'tik' at night and on cloudy days. Secretive, and difficult to find and catch.

Length: 2 cm ss: Snoring Puddle Frog

VINCENT CARRUTHERS

NG

Bubbling Kassina

Small stout frog with slender legs. The back may be fawn, yellow or olive, with distinctive **bold, dark stripes**. The sides are mottled and the underbelly is off-white. Shy and seldom seen. It occurs in grasslands in the vicinity of larger bodies of water such as dams and permanent pools. The call is a loud liquid 'quoip' made from the base of a grass tuft, often well away from water.

Length: 4 cm

VINCENT CARRUTHERS

NG   WR

Freshwater Fishes

With all of the streams draining north and south off the Witwatersrand having been negatively impacted upon by man's activities, the Johannesburg area has a sadly impoverished indigenous fish fauna.

Most fishes are difficult to identify, with slight differences in fin shape often being the only means of separation between species. In most cases, a fish must be caught before it can be identified. In this section only six of the more commonly encountered species are featured, and two of these are introduced aliens. Other fish which may be encountered include the Orange River Mudfish (south of the Witwatersrand) and, very rarely, the Madagascar Mottled Eel (north of the Witwatersrand) which breeds in the Indian Ocean. Alien Perch and Mosquito Fish (Top Minnow) occur in isolated populations.

Threats to the fish of the area include wetland draining, pollution from industry, mines and urban centres. Weirs, canals and dams divide river systems and may prohibit the migration of certain fish. Alien fish such as the Largemouth Bass may out-compete indigenous species. Invasive alien plants such as Kariba Weed *Salvinia molesta* may cover the water surface and block out light. Despite these negative factors, there is plenty of scope for fish conservation through anti-pollution measures, and the restoration of streams and other aquatic habitats.

Names used in this section follow those in *A Complete Guide to the Freshwater Fishes of Southern Africa* by Paul Skelton (Southern, 1993). This is a comprehensive and easy to use guide to the 245 indigenous and introduced freshwater fishes of southern Africa.

Threespot Barb

Small, sandy-coloured fish with silvery underside. The name is most apt, as it has **three dark spots** spaced more or less evenly down its side. Representative of one of the largest fish genera in Africa – *Barbus* – the minnows or barbs. Occurs in shoals in streams or open water, favouring the fringes where vegetation is present. Feeds on small insects and micro-fauna.

Length: 15 cm ss: Straightfin Barb; Goldie Barb

Smallmouth Yellowfish

Large, olive-bronze fish with off-white underparts, and **fins tinged with orange**. There are two pairs of tentacles (barbels) protruding from the upper lip. Another representative of the *Barbus* genus, but considerably larger than the previous species. Prefers clear water with a sandy or rocky bed, but is also found in some dams. A popular angling species. Omnivorous.

Length: up to 50 cm ss: Smallscale Yellowfish

Carp *

Very large, pale olive to bronze fish. Native to Asia, this species was first introduced to South Africa in 1859. It is now naturalised and widespread in still or slow-moving water. Feeds on a wide variety of plants and aquatic creatures. A favoured angling species, but it causes damage to aquatic food-chains. The popular Koi is an ornamental variation of the Carp.

Length: up to 80 cm SA record mass: 21 kg

R A JUBB

WR

Sharptooth Catfish (Barbel)

Very large, slate-grey fish with white underside. Better known by its alternate name of Barbel, the **compressed body**, bony head and thin tendrils (barbels) are characteristic. Fish of the *Clarias* genus are unique in being able to breath air. Favours slow-moving water where it hunts smaller fish, frogs and aquatic insects. May also scavenge from dead animals. Popular angling and food fish.

Length: up to 1.4 m SA record mass: 33 kg

R A JUBB

WR

Banded Tilapia (Bream)

Medium-sized, olive-brown fish with dark vertical bands on the flanks and often with a **dark spot on the dorsal fin**. Breeding males have the dorsal fin tipped in red. Occurs in a variety of aquatic habitats and suitable for garden ponds. Typical of the family, the pair bond is strong and eggs and hatchlings are guarded by the parents. Feeds on plant and animal matter.

Length: up to 23 cm

PAUL SKELTON

WR

Largemouth Bass *

Large, pale olive fish with dark bands. Native to North America, this is one of the world's most popular sport fishes, and was introduced to this country in 1928. The **deeply-cleft dorsal fin** is distinctive. Found in dams or slow-moving rivers, and favours clear water with floating vegetation. A predator of other fish, but also takes crabs and frogs – they are a hazard in natural ecosystems.

Length: up to 60 cm SA record mass: 4.5 kg

PAUL SKELTON

WR

73

Invertebrates

The animal kingdom is divided into two broad groupings. The 'higher' classes, from fishes to mammals, are typified by an internal skeleton and are known as **vertebrates**. The 'lower' classes lack an internal skeleton, are generally much smaller in size and are known as **invertebrates**. There is an enormous number of different invertebrates and new species are constantly being identified and named. Leaving aside the tiny protozoans and the various groups of worms, this section concerns itself only with some of the more conspicuous segmented animals – arthropods – as well as a brief look at the molluscs as represented by garden snails. With the exception of butterflies, **the invertebrates featured are described to family level only** and are therefore listed in plural. **Photographs depict typical members of the family**.

Despite their small size, many invertebrates are fascinating subjects to observe and they often allow a more intimate examination of their lives than do larger animals which actively evade people. All habitats, except those degraded and heavily polluted by man, support a wide variety of invertebrates and you will not need to venture far in search of small creatures to study – a flower patch in your garden, a rotting log, a solitary bush or a small pond are all certain to harbour a wide diversity. Invaluable tools to aid observation include a magnifying glass (8x or 10x is ideal), close-focussing binoculars, a fine-mesh net and a glass jar for temporarily housing specimens. More specific observation tips are provided in the accounts which follow. Insects and other invertebrates make excellent photographic subjects, but a macro lens is essential.

Snails, spiders and insects are considered pests by many people and the large number of deadly insecticides on the market is testimony to this. Although not advocating that invertebrates be allowed to overrun houses and gardens (some, such as fleas and mosquitoes, are harmful), it must be realised that these small creatures play a vital role in food chains. Any attempt to create a garden attractive to birds must take invertebrates into account, and toxic pesticides should be avoided. Due to their small size and the transportability of the eggs and larvae, invertebrates from one location may be easily – and inadvertently – introduced to another. Various species of alien snails and insects (including many from the Cape and other parts of the country) have established themselves in Johannesburg, and some of these have a subtle but serious impact on native plants and animals.

PETER LAWSON

Garden snails

Garden snails are small relatives of the octopus, and are typical molluscs – soft-bodied animals with no rigid skeleton or body segments. The majority are aquatic, with marine species such as oysters and mussels being among the best known. Snails secrete a brittle shell into which they retreat for protection. Most are herbivores, active after dark or in wet weather.

Length: up to 5 cm

River crabs

River crabs are primarily aquatic, but move readily over land. Eight legs are used for walking sideways and a front pair is modified into pincers which are used as eating utensils or weapons. The body and limbs are covered in a brittle shell, and the eyes set on short stalks. Favours rocky streams where carrion, small fish and tadpoles are eaten.
Shell width: up to 10 cm

PETER LAWSON

WR

Wood lice

Wood lice are the only crustaceans to have successfully colonised the land but are confined to cool damp places. Colonies may be common around houses where they live beneath planks, flower-pots, loose bricks or doormats. They feed mostly on decaying plant matter, but also on cardboard and paper. When threatened, they roll themselves into a tight ball. Immatures are white.
Length: 15 mm

A S SCHOEMAN

KC **G**

Centipedes

Representatives from four groups of centipedes occur in the area, with the type illustrated being the most familiar. All have paired **legs which spread out-ward** from the body segments, and move in a snake-like fashion. Colour varies from pale yellow to tan or even blue. Tail pincers are used in defence. Nocturnal predators of insects.
Length: up to 10 cm

PETER LAWSON

RF **S** **G**

Millipedes

Four groups of millipedes occur in the area, with the cylindrical type illustrated being the most familiar. Each body segment has two pairs of tiny, **hair-like legs which point downward**. Movement is slow but millipedes are avoided by most predators as they eject a pungent toxic fluid, and readily curl into a spiral if disturbed. Active by day, they feed mostly on decaying leaves and wood.
Length: up to 15 cm

NATIONAL PARKS BOARD OF S.A.

NG **S** **G**

Spiders and relatives

Spiders and their relatives are distinguished from other arthropods by having eight legs and no antennae. An additional pair of limbs, known as pedipalps, are set between the front legs and jaws. Most have eight eyes, the arrangement of which is a key to family identification. All spiders are predators and over-power prey – often insects larger than themselves – by injecting poison through their fangs. The great majority of spiders are completely harmless to man and play an important role in the control of flies, mosquitoes and other insects.

Spiders are extremely abundant in almost all habitats, and it is quite likely that one may even be watching you as you read this! Thousands of species occur in South Africa, and no attempt is made here (with the exception of the two dangerous widow or button spiders) to differentiate between members of the various families. Habitat symbols have been omitted due to lack of space. Only representatives from six familiar families are featured here, but once you have become acquainted with these, a variety of publications will facilitate more detailed study. *Southern African Spiders: An Identification Guide* by Martin Filmer (Struik, 1991) is the most comprehensive book. The Spider Club is a Johannesburg-based special-interest group (see p. 123).

Spiders may be divided into those which spin webs, and those which do not. Looking for webs is a sure-fire way of finding spiders, and getting out in the early morning to search dew-soaked vegetation is always rewarding. A mist-spray water bottle can be used to the same effect.

Wolf spiders
Small, long-legged spiders which chase prey on the ground. No web is made.

Orb-web spiders
Large-bodied spiders which ensnare prey in large hanging web. Male is tiny.

Widow (Button) spiders ☠
Small, round-bodied spiders. Venomous. Black Widow has **red stripe** on abdomen; Brown Widow has **red 'hour-glass'** shape.

Flower crab spiders
Small, triangular-bodied spiders which ambush prey in flowers. Colour matches that of the flower. No web is made.

Wall crab spiders

Small, flat spiders which run along rocks and walls in houses. No web.

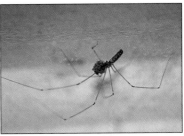

Daddy longlegs spiders

Tiny spiders with long, thin legs. Build sheet webs in the corners of rooms.

Rain (Wandering) spiders

Large, hairy spiders with long legs. Enter houses in summer. Eggs are in silk pouch.

Jumping spiders

Tiny, hairy spiders which pounce on insect prey. Most are black. No web.

Scorpions

Like spiders, scorpions have eight legs, but are easily identified by their large pincers and sting-tipped tail. Most are nocturnal, and all are predatory. Young are born live and ride on their mother's back. Two families occur in the region: the **Buthids** (illustrated) have small pincers and fat tails, and can deliver a painful (but rarely deadly) sting; the **Scorpionids** have large pincers and a smaller tail, and are less dangerous.

LEX HES

Ticks

Ticks are eight-legged, blood-sucking parasites of many animals including cattle, man and tortoises. Many transmit dangerous diseases such as 'tick bite fever' in man, and biliary in dogs. They will be familiar to all who have spent time walking in long grass in summer, when they are picked up as tiny 'pepper' ticks. They swell in size after gorging themselves on blood.

LEX HES

Insects

Insects may be told from all other arthropods in that they possess six legs in their adult stage. Furthermore, most insects are able to fly (on one or two pairs of membranous wings) – an attribute shared only by birds and bats. The great majority of insects, except for those which bite and sting, are overlooked by most people, yet insects are everywhere, with some 100 000 species estimated to occur in southern Africa. The small section which follows introduces only a few representatives from the more conspicuous families or groups. Butterflies – colourful, eye-catching and popular with naturalists – have been singled out for slightly more detailed treatment on pp. 84 to 87. The most interesting general publication on local insects is *African Insect Life*, written by S.H. Skaife and revised by John Ledger with superb photographs by Anthony Bannister (Struik, 1979). *Insects of Southern Africa* by C.H. Scholtz and E. Holm (eds) (Butterworths, 1985) is a detailed reference work, while *Pocket Guide: Insects* by E. Holm (Struik, 1986) is a useful introductory guide.

Insects undergo a series or physical changes as they grow, in a process known as **metamorphosis**. This may be 'incomplete' (egg, nymph, adult) as in grasshoppers, termites, dragonflies and others, or 'complete' (egg, larvae, pupa, adult) as in butterflies, mosquitoes, beetles, wasps and others.

Like most other invertebrates, insects are most active during the warm and wet summer months. In addition to stalking live insects (as discussed on p. 74) retrieving dead specimens from the surface of a swimming pool or off indoor window sills, and looking around an outdoor light, will reveal much of interest.

Dragonflies

Characterised by a helicopter shape (the aircraft was obviously modelled on this insect) and large eyes. The **wings are held at right angles to the body**. Adults catch insect prey on the wing, while the underwater nymph captures aquatic insects and small tadpoles. Found in the vicinity of water, where territorial males occupy perching posts. Mating adults fly in tandem.

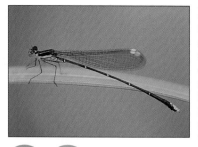

Damselflies

Generally smaller and more dainty than dragonflies. The **wings are held above the slender body** when at rest. The large eyes are more widely spaced than those of dragonflies. Adults catch tiny insects in flight, while the nymph (slender with leaf-like gills on its 'tail') captures aquatic insects. Usually found in the vicinity of water, but numbers may gather and rest in long grass in marshy places. Mating adults fly in tandem.

Termites

Herbivorous insects often referred to as 'white ants', but not related to ants at all. Colonies consist of individuals belonging to one of four castes – the reproductive 'queen' which may live for 20 years; sterile workers which collect food and construct nests; sterile soldiers which defend the colony; and winged reproductives (alates) which fly from the nest. Harvester termites have underground colonies, but most other families construct raised mounds.

LEX HES

G S NG

Praying mantids

Predatory insects with distinctive **enlarged forelimbs armed with spines** and used to capture flies and beetles. The mobile head is swivelled. All species rely on camouflage and range in colour from bright green to shades of pink. Females are larger than males and are said to sometimes eat their partner during copulation. A foamy white egg-sack is deposited in a sheltered place.

RF S G

Field crickets

The incessant chirping call of males is a familiar sound, but these herbivorous insects are nocturnal and seldom seen. The **antennae are very long**, and the powerful hind legs facilitate great leaps. One species – black with white patches at the base of its wings – may be a pest in the kitchens of homes. Close relatives are the green tree crickets and the furry, tunnel-dwelling mole crickets.

A S SCHOEMAN

NG PS G

King crickets

Nocturnal, predatory insects with a clockwork gait. The intimidating 'Parktown Prawn' often enters houses after dark and during wet weather. It is aggressive and armed with impressive jaws, but is unable to bite and relies on hissing as a form of defence. A foul black secretion is emitted under stress. Favours damp forests, but is now established in the well-wooded suburbs of Johannesburg.

PETER LAWSON

RF G

79

Shorthorned grasshoppers

Herbivorous insects with powerful hindlimbs usually held in a reversed V-shape above the cigar-shaped body. **The antennae are relatively short.** Flight is strong but usually only over short distances. Many species occur, with the Elegant Grasshopper (illustrated) being the most colourful. Most species are brown or green in colour. Locusts are gregarious, swarm-forming grasshoppers.

NG G S

Cockroaches

Flattened, oval-shaped insects which run rapidly in a scuttling motion. The antennae are thin and long, and the **folded forewings cover the entire body.** Rarely, if ever, takes to flight. One species – reddish-brown in colour – is common in many homes and storerooms around Johannesburg where it scavenges on plant and animal matter after dark. Other species occur under bark or stones.

G RF NG

Shield (Stink) bugs

The term 'bug' refers to a large and diverse group of insects characterised by **sucking mouthparts** rather than jaws. Most feed on plant juices and some – such as aphids – are serious pests. Shield bugs may be recognised by a triangular shield behind the head. A foul-smelling fluid is secreted in defence and has led to their alternate name. Other bug families are the cicadas, stainers and assassins.

G RF S

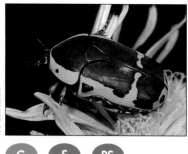

Fruit chafer (Rose) beetles

Two-thirds of all insects are classified as beetles of one kind or another. All are typified by **hard, shell-like outer wings** which protect the membranous flying wings. In contrast to bugs, beetles have **biting mouthparts.** The fruit chafers are not popular with gardeners as they feed on leaf buds and fruit. Fruiting peach trees may attract large numbers of these beetles which have the unnerving habit of flying directly at people.

G S PS

Ground beetles

Large flightless beetles which run on raised legs in pursuit of insect prey. Most are black and shiny with yellow or red markings. The jaws are powerful and can deliver a painful nip if the insect is handled. A pungent fluid may be squirted at attackers. Most ground beetles are nocturnal, but the species illustrated is often encountered on paths or in open areas during the day.

S NG

Blister beetles

Long slender beetles with leathery outer wings. The back is brightly-coloured in yellow, red or blue, while the head and legs are usually black. Adults feed by day on flower petals, but larvae are predators of grasshopper and bee eggs. Body fluids contain the drug cantharadin which is capable of causing blisters on human skin, and is believed to act as an aphrodisiac in some human cultures.

NG S G

Ladybird beetles

Small, brightly-coloured beetles with outer shells usually patterned in red and black. These striking colours warn potential attackers of their unpalatability. Most ladybirds are predators of aphids and scale bugs, and they are thus very popular with gardeners. Adults often congregate in autumn, settling in tightly-bunched swarms which spend the winter months hibernating in rock crevices.

RF NG G

Tenebrionid beetles

Most species of this group are wingless, and spend their time walking around slowly in rocky or sandy habitats. The species illustrated is among those better known as 'tok-tokkies' – a name derived from their habit of tapping the ground with their abdomen as a means of attracting a mate. Adults and larvae feed primarily on leaf litter, fungi and seeds. Over 3 000 species are known to occur in southern Africa.

S KC NG

81

Mosquitoes

Slender relatives of the flies, mosquitoes are best known for their blood-sucking habits. Most are active at night. Only the females suck blood – as a prelude to egg laying – and make the irritating whining noise. Males are silent, feed on plant sap, and have feathery antennae. Eggs are laid in still water, where the larvae develop. May transmit malaria and yellow fever, but not in the Johannesburg area.

WR RF G

Flies

A vast number of different fly species occur in the region. All are characterised by a single pair of translucent wings and sucking or licking mouthparts. The larvae are legless, and known as maggots. Among the more conspicuous flies in the region are the muscid (house) flies, horse flies, fruit flies and blow flies. House flies feed on manure and rotting plants, and transmit various diseases.

G PS NG

LEX HES

Honey bees

Important pollinators of flowering plants, honey bees are very sociable and form huge swarms centered around a single, egg-laying 'queen'. The colony consists of workers – which make the multi-celled honey combs – and drones. Hollow trees are favourite nest sites, but house eaves, and nest boxes intended for birds, are also used. Bees sting as a last resort, as they die soon after.

G PS KC

LEX HES

Ants

Ants are considered to be the most advanced of social insects. Colonies consist of a 'queen', workers and soldiers. Males and dispersing females are winged. Food varies according to species, but most are carnivorous. Several different families occur, including the large ponerine ants, the driver (or red) ants, and the house ants. The Argentine Ant is an alien invader which attacks other ant species. Some ants 'milk' aphids for the 'honeydew' they produce.

G NG S

Parasitic wasps

A sizeable family of slender wasps characterised by the very long ovipositor of the female. Eggs are laid directly into the living larvae of a moth, butterfly, bee or other insect. Some species bore through the ground, or through wood, in order to reach a host. Developing wasp larvae feed on body fluids of the host. Many species are brightly-coloured, and most do not possess stings.

RF　G　S

Paper (Social) wasps

These wasps construct papery nests, made of finely-chewed bark, and hung by a single stalk from overhangs such as eaves, ceilings and caves. Nests are aggressively defended and adults deliver a painful sting. Species of the genus *Belonogaster* are large with thin waists, while those of *Polistes* are smaller with a striped abdomen. Nectar is the main food but larvae are fed on caterpillars.

KC　S　G

Moths

With their scale-covered wings, moths belong to the same group as butterflies but are mostly nocturnal, and attract less interest from lepidopterists. Over twenty different families occur in the area, with hundreds of different species. Some, such as the Emperor and Hawk moths, are large and impressive, but most are small and insignificant. Differences between moths and butterflies are subject to several exceptions, but in general, moths have **feathery or tapered antennae,** as opposed to the club-tipped antennae of butterflies. Most day-flying moths are brightly-coloured and may be confused with butterflies. Moths navigate by starlight but are disorientated by artificial light; large numbers are attracted to outdoor lights. Adults feed on nectar. The larvae (caterpillars) feed on leaves. Most larvae are hairy and spin cocoons of silk. *Moths of Southern Africa* by E. Pinhey (Tafelberg, 1975) is the standard reference book on these insects.

RF　S　G

Butterflies

Because of the interest they hold for naturalists, butterflies are treated here in greater detail than other insect groups. Even so, only twelve of the most common or interesting species are featured. The aim of this section is not only to enable you to identify these common types, but also to introduce the major families which will help you to identify other butterflies to this level.

The general structure and biology of butterflies is closely related to that of moths, with the most obvious butterfly traits being day-flying habits and clubbed antennae. The pupae (chrysalis) of butterflies differ from those of moths in being naked. Adults are nectar or sap feeders, and many drink from puddles or moist soil. Eggs are deposited on the underside of leaves which the larvae (caterpillars) then feed upon; most species have specific food plants.

Watching butterflies is not difficult, but certain strategies will improve your chances of approaching these often restless insects. Butterflies have good eyesight, so walk slowly towards them, whilst keeping a low profile against the skyline. Avoid casting your shadow over a butterfly, as this is sure to scare it off. A bait of fermenting bananas may serve to attract many species.

There are several books on local butterflies, with *Pennington's Butterflies of Southern Africa*, revised by Pringle, Henning & Ball (Struik, 1994), being the definitive reference work. *Butterflies of Southern Africa – A Field Guide* by Mark Williams (Southern, 1994), is an excellent and easy to use guide which features 233 of the more than 800 local species; names used in the following accounts follow those used in this book.

African Monarch

Conspicuous, coppery-red butterfly which flies low, in a leisurely fashion. Sexes differ in the number of black spots on hindwing. They are unpalatable to predators due to toxins absorbed by the larvae feeding on Milkweed *Asclepias fruticosa*, and are mimicked by several palatable species including the female Common Diadem (below). Fairly easy to approach when feeding at flowers.

Common Diadem

The distinctive pied colouration of the male renders it easily identifiable, but the female is an accurate mimic of the unpalatable African Monarch (above), differing in having only a single black spot on each hindwing. An active flyer, usually at a low height. Adults feed from a variety of flowering plants. The eggs are laid on succulent plants belonging to the Portulacaceae family.

Painted Lady

Paler and more spotted than the African Monarch (and its mimics), this butterfly is tolerant of a wide variety of climatic zones and habitats. The eggs are laid on a wide variety of food plants including members of the Asteraceae and Fabaceae families. Flight is rapid and erratic, but males often pause to display on exposed ground. One of the few butterflies active on cold winter days.

S   NG   G

Yellow Pansy

Pale, yellow and black butterfly with distinctive **blue wing spots**. It flies with a characteristic double wing beat followed by a glide on outstretched wings. Territorial males perch with open wings on bare ground, but fold them up if disturbed. Adults feed on nectar and fluid from herbivore dung, and regularly drink from puddles. The eggs are laid on plants of the Acanthaceae family.

LEX HES

S   NG   G

Foxy Charaxes

Robust butterfly with beautifully-patterned underwings, and with a pair of thin 'tails' on each hindwing. Flight is fast and direct. Males establish territories on hilltops and perch conspicuously on the edge of a bush or tree. Mostly restricted to undisturbed habitats, rarely entering gardens or parks. Adults feed on fermenting fruit and carnivore dung. Eggs are laid on a variety of trees.

STEPHEN WOODHALL

S   KC

Garden Acraea

Long-winged butterfly with partially transparent forewings. The flight is slow and leisurely, and usually close to the ground. The adults and caterpillars are apparently toxic, and avoided by birds. Adults feed on nectar and are frequently seen in gardens. The eggs are laid on the leaves of the Wild Peach *Kiggelaria africana* (p. 97) and the emerging caterpillars may defoliate the whole tree.

AS SCHOEMAN

S   RF   G

Eyed Bush Brown

Small, dark brown butterfly with distinctive 'eye' patterns on the upper and lower wings. Favours shady areas, and is often seen among long grass along rivers. The flight is slow, lazy and erratic. Frequently perches with folded wings, often among dry leaves where it is well camouflaged. Adults feed mostly on the sap of fallen fruit. The eggs are laid on various species of grass.

RF WR G

STEPHEN WOODHALL

Citrus Swallowtail

Large, pale yellow butterfly with black patterning. As a group, the swallowtails are typified by tail-like protuberances from the hindwing, but this, the most common species, lacks these 'tails'. It flies purposefully at medium height, and often settles to feed on flowers. Males hold territories in bush clumps and gardens. Eggs are laid on the Small Knobwood *Zanthoxylum capense* (p. 93) and other plants of the Rutaceae family.

RF G S

STEPHEN WOODHALL

Patricia Blue

Tiny butterfly belonging to a large group of similar species known as 'blues'. The male is pale blue on the upperwings, and creamy-white below. The female has brown edging to the wings. Both sexes have a small black spot, edged in gold, on the lower wings. Feeds on nectar and often drinks at puddles. The flight is erratic, and usually low. Eggs are laid on *Salvia* and *Lantana* species.

G NG S

Common Hottentot

Small, drab and moth-like butterfly with short wings, hairy body and large eyes. Other butterflies in this family include the 'skippers' and 'policemen'. Usually found in areas with short grass where it is most active at dusk. Flies rapidly at a low height, often settling in a distinctive way with only the forewings folded upwards. Adults feed on nectar. The eggs are laid on various grasses.

NG S G

Brownveined White

Small, mostly white butterfly with chocolate-brown patterns on the wing tips and a yellow wash underwing. This is the butterfly which migrates in great numbers – in a north-easterly direction – during late summer and autumn. When not migrating, it flies erratically, feeding on nectar and drinking from puddles. Eggs are laid on *Maerua* and *Boscia* trees – both rare in the Johannesburg area.

G   PS   NG

African Migrant

About twice the size of the previous species, the males are white above with a pale green underwing. The females are lemon-yellow above with tan wing tips, and are mottled yellow below. May form large flocks which migrate in summer or autumn, often in the company of the Brownveined White. Feeds on nectar and often visits puddles. Eggs are laid on various species of *Cassia*.

WILDERNESS SAFARIS/COLIN BELL

G   PS   NG

Common Orange Tip

Similar in appearance and size to the Brownveined White, the males are snow-white with orange-yellow tips to the forewings and a brown smudge between the wings. The female differs in having less white above. Males tend to fly more rapidly and purposefully than females. Active only in sunny weather, they are especially attracted to purple flowers. Eggs are laid on *Capparis* species.

NG   G   S

Broadbordered Grass Yellow

Small, lemon-yellow butterfly with brown borders to tips of the fore and hind wings. The sexes are similar. The flight is weak and erratic and usually close to the ground. Flower nectar is the food, and these butterflies may be common in gardens where they flutter across lawns. Males visit puddles and damp soil to suck up moisture. Eggs are laid on low-growing herbs.

STEPHEN WOODHALL

NG   PS   G

Trees and Woody Shrubs

A surprising variety of indigenous trees occur in the Johannesburg area, with the greatest diversity on and around rocky ridges. In addition, large numbers of alien species are grown in gardens, parks and along streets, and some of these have become undesirable and troublesome invaders which should be eradicated wherever possible. This section features 58 of the more conspicuous or common trees and woody shrubs – both indigenous and alien – which propagate themselves. Trees such as oaks and planes – which are widely cultivated, but which do not reproduce themselves in the Johannesburg area – are not featured.

The identification of trees and shrubs is considered a daunting prospect by beginners, but one way of getting to know them is **go out and look for the species featured here**. Once you have become familiar with these plants, you will be able to recognise additional, less common trees, and will be ready to take your study further. A number of detailed books on trees and shrubs are available, and these are listed on p. 122. The most comprehensive book is *Trees of Southern Africa* by Keith Coates Palgrave (Struik, 1983), but this does not cover alien species in any detail.

In this section, the **scientific names** of the trees are given before the common names. It is wise to learn and use these names, as any future study of trees will involve comparison between related species. Family names are also included.

Many of the indigenous species featured here are rewarding garden subjects as they attract an abundance of insects and birds, and are adapted to local climatic conditions. Invasive aliens (marked *) should not be grown in gardens.

Aloe marlothii
Mountain Aloe/Bergaalwyn

Spiny succulent with fleshy, blue-green leaves covered in prickles. The stem is unbranched, bearded with old dry leaves. A branched panicle, with horizontal arms, holds the orange-red flowers. A familiar sight on hill slopes and most striking when in flower from June to August. Copious nectar attracts insects, sunbirds and other birds.

ASPHODELACEAE Height: 2 to 5 m

Agave americana *
Sisal

Spiny succulent with fleshy leaves tipped with spikes. Leaves may be grey-blue or green and yellow. **Stemless**, but produce a massive woody flower stalk after about 10 years, after which the plant dies.

A **native of central America**, but widely cultivated in gardens and often used as a hedge on farms. Now an escapee in the vicinity of settlements where it reproduces by means of suckers.

AGAVACEAE Height: up to 6 m (incl. flower)

Salix babylonica *
Weeping Willow

Large drooping tree with narrow leaves.
Grows only near water, and is often
cultivated in parks. A probable **native of
China**, it is one of the most familiar
trees. Only female plants occur in South
Africa, but the species reproduces itself
vegetatively, as fallen branches take
root. The soft branches are favoured by
hole-nesting birds. Deciduous.
SALICACEAE Height: up to 10 m

PS RF WR

Salix mucronata
Vaal Willow/Vaalwilger

Small slender tree related to the alien
Weeping Willow. Grows on the banks of
streams. Often multi-stemmed from the
base with a bushy, hanging shape being
typical. Narrow leaves are very similar
to those of *S. babylonica*: alternate and
finely serrated on the margins. Spikes of
green-yellow flowers appear in summer.
Deciduous.
SALICACEAE Height: 2 to3 m

WR

Populus canescens *
Grey Poplar

Slender upright shape and smooth **pale
bark**. Leaves are round with entire margins
on larger trees, triangular and toothed on
saplings; pale grey of underside is obvious
in windy conditions. Invasive alien from
Europe, it **forms groves in moist places**.
Rapidly colonises suitable terrain,
destroying wetlands and indigenous
flora in the process. Deciduous.
SALICACEAE Height: up to 15 m

PS WR

Celtis africana
White Stinkwood/Witstinkhout

Handsome spreading tree with **smooth,
pale grey trunk**, sometimes buttressed at
the base. Leaves are heart-shaped, finely
toothed and alternate; soft and lime-green
in spring, then leathery and dark green in
summer before turning yellow. Flowers
inconspicuous. Fast growing and an ideal
garden subject. White trunk and branches
of dead trees are conspicuous. Deciduous.
ULMACEAE Height: up to 20 m

PS S RF

Morus alba *
White Mulberry

Large bushy tree with drooping shape. The large, shiny, oval-shaped leaves are strongly toothed and terminate in a fine point. A **native of Asia**, this popular garden subject is now an invader along streams. The purple, raspberry-like fruits appear in early summer and are relished by birds which act as agents for its spread. Deciduous.

MORACEAE Height: up to 10 m

Ficus ingens
Redleaved Fig/Rooiblaarvy

Small, low-growing tree which favours rock crevices. Pale grey bark is smooth. The alternate leaves are leathery, oval and pointed. Most conspicuous in spring when the new foliage is bright coppery-red in colour. Small figs appear in summer and are eaten by various birds. It may attain a much larger size in warmer regions. Deciduous.

MORACEAE Height: 1 to 3 m

Protea caffra
Common Sugarbush/Suikerbos

Short rounded tree characterised by its leathery, **blue-grey leaves** and **rough black trunk**. Forms colonies on hillsides, especially abundant in the Klipriviersberg and along the Witwatersrand. Attractive flowers appear in summer. Close relatives – *P. roupelliae* and *P. welwitschii* – occur in a few isolated pockets on the Witwatersrand.

PROTEACEAE Height: 3 to 4 m

Grevillea robusta *
Australian Silky Oak

Tall straggly tree with many bare branches and an untidy appearance. Most striking in early summer when the sprays of golden-yellow flowers stand out against the grey-green foliage. **Leaves are fern-like in shape, with pale undersides**. Fruits are woody capsules. Cultivated in gardens and parks, this **alien** is now naturalised in the area.

PROTEACEAE Height: up to 15 m

Mundulea sericea
Corkbush/Kurkbos
Well-proportioned shrub or small tree
which grows mostly on hillsides or at the
base of koppies. Fire-resistant. The bark
is corky, deeply furrowed, and yellowish-
brown (black in burnt trees). **Foliage is
silvery-green.** Compound leaves are
pale underneath. Mauve, pea-shaped
flowers appear in midsummer, and the
velvety pods are sandy-brown. Evergreen.
FABACEAE Height: 1 to 2 m

Sesbania punicea *
Brazilian Glory Pea ☠
Slender, multi-stemmed shrub with
droopy compound leaves and smooth
bark. A **native of South America**, it now
colonises stream banks. Most obvious in
summer when the **bright orange, pea-
shaped flowers** cover the bush. The pods
are distinctive, four-winged capsules.
Seeds, leaves and flowers are highly
poisonous. Deciduous.
FABACEAE Height: 2 to 4 m

Crotalaria agatiflora *
Bird Flower
Slender, multi-stemmed shrub or small
tree with drooping shape. A **native of
East Africa**, but now an invasive weed of
disturbed soils, particularly along roads.
The **trifoliate leaves are grey-green.**
Brilliant lime-yellow flowers hang in
clusters at the stem tips, and are said to
resemble a row of perched birds. Flowers
during summer and autumn.
FABACEAE Height: 1 to 3 m

Erythrina zeyheri
Ground Coral Bush/Ploegbreker
Small shrublet with a substantial woody
rootstock (hence the Afrikaans name
'plough-breaker'). The magnificent
scarlet flowers appear in spring and are
followed by constricted pods housing
large red 'lucky-beans'. The large
trifoliate leaves are covered with small
hooked spines. Usually grows in colonies
in clay soils near wetlands. Deciduous.
FABACEAE Height: 50 cm

91

Acacia caffra
Common Hookthorn/Haakdoring
Graceful gnarled tree typical of hillsides. Compound leaves are droopy and fern-like. Bark is dark brown, rough and flaky. **Dense spikes of cream flowers** are profuse in spring, often appearing before the leaves. Pods are narrow. **Small hooked thorns are paired**, most prevalent on new growth. The similar *A. ataxacantha* has unpaired thorns.
MIMOSACEAE Height: 3 to 8 m

Acacia karroo
Sweet Thorn/Soetdoring
Rounded tree usually growing in clay soils. Bright green, compound leaves become dull green in autumn. Bark of older trees is dark and rough, that of saplings is reddish and smooth. Straight **white thorns are in pairs**, most abundant on new growth. Flowers are sweetly-scented, **golden-yellow balls** in summer. Pods are sickle-shaped. Deciduous.
MIMOSACEAE Height: 3 to 8 m

Acacia podalyriifolia *
Pearl Acacia
Small bushy tree with distinctive, **silvery-grey foliage** and **no thorns**. A **native of Australia**, this decorative tree is widely cultivated, but is now an invader of open spaces and roadsides. The flowers are **golden-yellow balls** which appear in profusion in autumn. The leaves are oval and pointed. Evergreen. The similar *A. baileyana* * has tiny, compound leaves.
MIMOSACEAE Height: 3 to 4 m

Acacia mearnsii *
Black Wattle
Bushy, much-branched tree with an untidy appearance. A **native of Australia**, it has spread over much of the country where it forms dense thickets along streams and on hillsides. Similar to *A. karroo*, but **lacks thorns**. Compound leaves are blue-green, the flowers are sweetly-scented, **pale yellow balls**. Pods are flat and constricted between seeds. The similar *A. dealbata* * has grey leaves.
MIMOSACEAE Height: 3 to 5 m

Zanthoxylum capense
Small Knobwood/Perdepram

Small slender tree with smooth grey trunk and stems covered in small spines. The compound leaves are unusual in that the **paired leaflets become progressively larger towards the tip**; glossy and lemon-scented. Tiny yellowish flowers are in clusters. Fruits resemble miniature lemons. Often grows with other species in thickets. Deciduous.

RUTACEAE Height: 2 to 4 m

Lannea discolor
Livelong Tree/Dikbas

Small, open-branched tree. Usually the first to shed its leaves in autumn and last to show new growth in spring. The foliage is distinctive, with the compound leaves being soft and pale **matt green above and silvery-white** below. Leaves turn a beautiful rusty-red in autumn. The bark is smooth, pale grey. Tiny flowers appear before new leaves in spring.

ANACARDIACEAE Height: 2 to 4 m

Ozoroa paniculosa
Common Resin Tree/Harpuisboom

Gnarled, irregularly-shaped tree usually growing in small colonies on hillsides. Hardy and fire-resistant. The bark is thick and rough. Oval, silvery-green leaves have **distinctive parallel veins** and a pale undersurface. Flowers are inconspicuous. Small, raisin-like fruits are white, ripening to black. Stems exude a resinous sap. Evergreen.

ANACARDIACEAE Height: 3 to 5 m

Melia azederach *
Syringa/Persian Lilac

Tall graceful tree with glossy, fern-like, compound leaves. A **native of south-east Asia**, it is a vigorous invader, particularly along streams. Small lilac flowers appear in dense clusters in spring, and are very fragrant. Yellow berries are produced in abundance in autumn – poisonous to man, but eaten by birds. Should not be grown in gardens or parks. Deciduous.

MELIACEAE Height: up to 20 m

93

Rhus dentata
Nanaberry/Nanabessie

Small willowy shrub often growing beneath larger trees. The **trifoliate leaves** – characteristic of the *Rhus* genus – are **broad and strongly toothed**; dull or glossy green, turning to shades of yellow, red and gold in autumn. Flowers are inconspicuous. Small, yellow-to-red berries are in grape-like bunches. Decorative garden subject. Deciduous.
ANACARDIACEAE Height: 1 to 3 m

Rhus lancea
Karree/Karee

Graceful willowy tree with twisted shape. Crooked branches emerge near the base. Bark is dark brown and rough. **Trifoliate leaves** glossy, dark green, slender with **no serrations**. Flowers are inconspicuous. Small **round berries** hang in dense bunches. Fire and drought-resistant. Good, fast-growing garden subject. Evergreen.
ANACARDIACEAE Height: 3 to 7 m

Rhus leptodictya
Mountain Karree/Bergkaree

Graceful willowy tree with twisted shape. Crooked branches emerge near the base. Bark is dark brown and rough, but young stems are reddish-brown. **Trifoliate leaves** pale or greyish-green, slender and **distinctly serrated**. Flowers are inconspicuous. Small, **flattened, box-like berries** hang in bunches. Good, fast-growing garden subject. Evergreen.
ANACARDIACEAE Height: 3 to 6 m

Rhus pyroides
Common Wild Currant/Taaibos

Small bushy shrub with smooth, dark brown trunk. **Trifoliate leaves** are velvety, pale green. The **stems and twigs are armed with spines**. Often colonises disturbed soils and termite mounds; once established, saplings of other species often grow in its shade, and a bush clump forms. Flowers and fruits similar to those of *R. dentata* and *R. lancea*. Deciduous.
ANACARDIACEAE Height: 2 to 5 m

Maytenus heterophylla
Common Spikethorn/Pendoring

Bushy shrub or small tree with a straggly rigid shape. Bark is dark brown, deeply furrowed and fire-resistant. **Branches and stems are armed with strong, straight spines**. Small, leathery leaves are mid-green or blue-green. Cream-coloured flowers mass over the whole tree in September but are foul smelling. Fruit is a three-lobed capsule. Evergreen.
CELASTRACEAE Height: 2 to 4 m

Pappea capensis
Jacketplum/Doppruim

Gnarled, open-branched tree usually growing among rocks or on termite mounds. Bark is grey or pale brown. **Oval leaves are parchment-like with characteristic wavy margins; often serrated**. Small cream flowers appear in late summer, attracting many insects. Fruit is a small brown capsule enclosing delicious red jelly. Deciduous.
SAPINDACEAE Height: 2 to 7 m

Ziziphus mucronata
Buffalothorn/Wag-'n-bietjie

Droopy, densely-foliaged and drought-resistant tree. Bark is smooth at first, but becomes rough with age. Alternate leaves are oval, serrated and **glossy** on their upper surface. Small but prolific **thorns are arranged in pairs of one straight, and one decurved**. Tiny green flowers appear in summer; followed by round, brick-red berries. Deciduous.
RHAMNACEAE Height: 2 to 7 m

Rhamnus prinoides
Dogwood/Blinkblaar

Small shrub with dense, **very glossy foliage**. Smooth, grey-brown trunk. Shiny leaves are dark green above, pale and dull below; simple, alternate with slightly quilted appearance and finely serrated. Flowers are inconspicuous. Small berries turn from yellow to red, then black; carried in profusion and loved by birds. Rewarding garden subject. Evergreen.
RHAMNACEAE Height: 1 to 3 m

95

Leucosidea sericea
Oldwood/Ouhout
Small bushy shrub or small tree which grows along streams and in sheltered kloofs. The **gnarled shape and rough, reddish-brown bark** are distinctive. Young stems are hairy. Foliage is grey-ish-green. The compound leaves are hairy, pale underneath, aromatic and shaped like those of a rose. Cream flowers appear in spring. Evergreen.
ROSACEAE Height: 2 to 4 m

Jacaranda mimosifolia *
Jacaranda
Handsome, open-branched tree with a broad crown. A native of South America, it is widely cultivated along streets and in parks and gardens, but invades natural habitats. Unmistakable when mauve blossom covers the tree in early summer. Leaves are compound, fern-like. Fruits are flat woody capsules housing winged seeds.
BIGNONIACEAE Height: 5 to 20 m

Eucalyptus spp *
Gums/Eucalypts
Tall, open-branched trees **native to Australia**. Cultivated in plantations for wood and pulp. Several species are now naturalised in the Johannesburg area. Often grown as wind-breaks. Most species have distinctive bark which peels off to reveal pale stems and trunk. Leaves are usually lance-shaped. Fruits are woody capsules. Pollen-rich flowers attract bees. Evergreen.
MYRTACEAE Height: up to 50 m

Rhoicissus tridentata
Bushman's Grape/Boesmansdruif
Sprawling shrub which may gain height by clinging to other plants. Could be confused with *Rhus dentata* (p. 94), but the toothed, trifoliate leaves are more rounded, with **asymmetrical side leaflets** and thin tendrils. Leaves turn red in autumn. A relative of the true grape, the large berries are eaten by birds and porcupines.
VITACEAE Height: 1 to 3 m

Grewia occidentalis
Crossberry/Kruisbessie

Drooping shrub with a tendency to scramble, often growing in bush clumps with other species. Leaves are simple, alternate, thin and finely serrated; light green above, paler below. **Star-shaped flowers are mauve with yellow centres**. Fruits are reddish, four-lobed capsules, resembling glued-together berries. Attractive garden subject. Deciduous.
TILIACEAE Height: 2 to 4 m

Dombeya rotundifolia
Common Wild Pear/Drolpeer

Crooked bushy tree typical of hillsides and conspicuous at the end of winter when it is covered in masses of **papery white flowers** (similar to those of the true pear tree). Bark is dark and furrowed. Roundish leaves are alternate, irregularly toothed and coarsely textured. Fruits are inconspicuous capsules. Fire-resistant and hardy. Deciduous.
STERCULIACEAE Height: 3 to 5 m

Kiggelaria africana
Wild Peach/Wildeperske

Small to medium-sized tree often found in bush clumps or along streams. Trunk is smooth, becoming rougher with age. Simple leaves are oval, alternate, dull green above, paler and velvety below. Leaf shape is variable, with those on new growth usually serrated. Flowers inconspicuous. Fruit is a round leathery capsule, splitting to release orange seeds. Evergreen.
FLACOURTIACEAE Height: 3 to 7 m

Combretum erythrophyllum
River Bushwillow/Vaderlandswilg

Tall, open-branched tree often dominant in riverine forest. Bark is smooth, pale brown and mottled. Simple leaves are opposite, narrowly oval, with pointed tips. Leaves turn rust-red in autumn. Cream flowers are inconspicuous, but small, **four-winged fruits** are distinctive. The related *C. molle* has larger, velvety leaves and grows on dry hillsides.
COMBRETACEAE Height: up to 13 m

Cussonia paniculata
Mountain Cabbage Tree/Kiepersol

Distinctive, lollipop-shaped tree topped with a mop of blue-green leaves. **Usually single-stemmed**. Bark is dark brown, deeply furrowed, corky and fire-resistant. Leathery leaves are very large, fan-shaped with seven to nine leaflets. Small greenish flowers and fleshy fruits are held in clusters on branched stems. Interesting garden subject. Evergreen.
ARALIACEAE Height: 3 to 6 m

Heteromorpha trifoliata
Parsley Tree/Pietersieliebos

Small spindly tree with sparse foliage and bare stems. The distinctive bark is smooth and shiny, peeling off in papery sheets to reveal a bright coppery under-layer; older branches are **ringed in bamboo fashion**. Compound leaves are variable, with 3, 4, 5 or 7 leaflets. Flowers and fruits are in sprays. Interesting garden subject. Deciduous.
APIACEAE Height: 2 to 6 m

Englerophytum (=Bequaertiodendron) *magalismontanum*
Transvaal Milkplum/Stamvrug

Dense shrub with dark green foliage. Leaves are shiny, leathery and clustered at end of stems; underside covered in fine, rusty-brown hairs. New growth is silvery-green. Small flowers have a musky odour. Edible **red fruits grow on the trunk and stems**. All parts exude a milky latex. Evergreen.
SAPOTACEAE Height: 1 to 3 m

Euclea crispa
Blue Guarri/Bloughwarrie

Dense shrub or small tree, with trunk and stems not usually visible. **Foliage is dull, blue-grey or green**. Simple leaves are opposite, rigid, leathery and point upwards. Leaves of individual plants may vary in shape and colour. Flowers are inconspicuous. The fruits are pea-sized berries. The bark is mottled, becoming rough with age. Very hardy – resistant to frost, drought, and fire. Evergreen.
EBENACEAE Height: 2 to 6 m

Diospyros lycioides
Transvaal Bluebush/Bloubos

Rounded shrub, or small straggly tree, typical of bush clumps and often forming groves. Multi-stemmed, with smooth bark. **Oval leaves are hairy**, dull green and clustered at the ends of branches; the **veins are prominent**. Small, cream, bell-shaped flowers appear in spring. Fruits resemble tiny tomatoes and are eaten by birds and dassies. Evergreen.
EBENACEAE Height: 2 to 4 m

Diospyros whyteana
Bladdernut/Swartbas

Bushy shrub or small tree with a drooping shape. Oval leaves are alternate, dark green, shiny and with distinctive soft hairs on margins. **Orange or red leaves often persist on the stems** and stand out amongst the greenery. Flowers are cream-white bells. Fruits are held in papery capsules, reddish when ripe. Decorative garden subject. Evergreen.
EBENACEAE Height: 1 to 3 m

Olea europaea subsp. *africana*
Wild Olive/Olienhout

Medium-sized, dense tree with a rounded shape. Trunk is gnarled with dark, rough bark. Simple leaves are slender, opposite, leathery, paler below, and with pointed tips; the **veins are indistinct**. Flowers are inconspicuous. Fruit is a small, round or oval berry, purplish when ripe. Hardy and long-lived, this makes a good screen in the garden. Evergreen.
OLEACEAE Height: 3 to 7 m

Buddleja saligna
False Olive/Witolienhout

Small straggly tree with foliage very similar to the Wild Olive. Simple leaves are slender, opposite, paler below and with pointed tips; the **veins are prominent** on the under-surface. Bark of older trees is stringy and fawn-coloured. Tiny flowers are in showy heads in early summer. Fruits are inconspicuous brown capsules. Evergreen.
LOGANIACEAE Height: 3 to 6 m

Buddleja salviifolia
Sagewood/Saliehout
Graceful shrub with drooping, willowy shape. Bark is grey-brown and stringy. Leaves are dull grey-green, **heavily crinkled above, velvety and white below**; veins are prominent and the margins are finely serrated. Small lilac flowers appear in large profuse heads in midwinter; sweetly scented. Decorative garden plant. Evergreen.
LOGANIACEAE Height: 2 to 4 m

Solanum mauritianum *
Bugweed Tree
Open-branched shrub or small tree with **large, velvety leaves**; dull green above and pale below. A **native of South America** but now an invasive alien most prolific on disturbed soils and along streams. Mauve, star-shaped flowers are in upright clusters. Round berries are favoured by many birds. Crushed leaves are foul-smelling.
SOLANACEAE Height: 3 to 5 m

Ehretia rigida
Puzzlebush/Deurmekaarbos
Tangled shrub or small tree with an intricate drooping shape. The stem is smooth grey-brown. Small leaves are clustered on short stems; leathery but variable in texture and shape. Similar to the Common Spikethorn (p. 95) but **lacks spines**. Pale lilac flowers are sweetly scented and borne in profusion in spring. Berries green, ripening to red. Deciduous.
BORAGINACEAE Height: 2 to 4 m

Halleria lucida
Tree Fuchsia/Notsung
Drooping shrub or small tree with rough, flaky bark. Simple leaves are opposite, shiny, finely serrated and have pointed tips; bright green, turning purple in autumn. Tubular **orange flowers grow on stems** and are often screened by foliage; rich in nectar and loved by sunbirds. Berries green, ripening to black. A fine garden subject. Deciduous.
SCROPHULARIACEAE Height: 2 to 5 m

Rothmannia capensis
Cape Gardenia/Katjiepiering

Bushy shrub or upright tree with **layered branching**. Bark of mature plants has the appearance of hessian. Simple leaves are oval, opposite, glossy, leathery; dark green above, paler below. Large, white, trumpet-shaped flowers have pink spots inside; sweetly scented. Fruit is a large, green drupe. Decorative garden subject, but slow growing. Evergreen.

RUBIACEAE Height: 2 to 5 m

Vangueria infausta
Wild Medlar/Wildemispel

Small tree or open-branched shrub with an unkempt appearance. Knobbly stem is smooth. Simple leaves are opposite, oval, large, furry, and **often covered with raised lumps** caused by gall insects. Older leaves are rough and brittle before dropping in autumn. Flowers are small, greenish-white. Edible round fruit is green, ripening to yellow. Deciduous.

RUBIACEAE Height: 3 to 4 m

Canthium gilfillanii
Velvet Rock Alder/Fluweelklipels

Small tree with layered **branches and stems growing at right angles**. The small oval leaves are velvety in texture, often with the margins rolled under. The pale grey bark is smooth. The creamy-green flowers are inconspicuous. The fruit is a small oval berry, black when ripe. Grows on stony hillsides, often in colonies. Deciduous.

RUBIACEAE Height: 2 to 3 m

Landolphia capensis
Wild Apricot/Wildeappelkoos

Dense, low-growing shrub with **trailing branches**. The small oval leaves are glossy green above, paler below with distinct veins. The **white flowers are star-shaped** and jasmine-scented. The round, golf-ball-sized fruits are orange when ripe. The skin is leathery, but the fleshy pulp is edible. All parts contain a **milky latex**. Evergreen.

APOCYNACEAE Height: up to 1 m

Tarchonanthus camphoratus
Wild Camphor Bush/Saliehout

Small tree or shrub with a drooping shape. The main stem is gnarled and twisted and the bark is furrowed on older trees. Simple leaves are alternate, grey-green above, paler below. Female trees bear tiny flowers, followed by **white, cotton-like balls** which contain the seeds. Very showy in autumn and winter. Evergreen.
ASTERACEAE Height: 2 to 4 m

Brachylaena rotundata
Mountain Silver Oak/Bergvaalbos

Tall, open-branched tree or bushy shrub, often growing in exposed positions among rocks. Bark is dark and deeply ribbed. **Foliage is silvery-grey**. Simple leaves are alternate, crinkly, margin scalloped and toothed; grey-green above, silvery-white below. Small daisy flowers are in massed heads in spring. Fruits inconspicuous.
ASTERACEAE Height: 3 to 8 m

Nuxia congesta
Wild Elder/Wildevlier

Small bushy shrub or tree with pale stringy bark, peeling off in strips. The **stems are distinctively four-angled**. Leaves are in **whorls of three**, and clustered at the ends of branches; variable in shape, they may be shiny or hairy, with margins toothed or entire. Tiny, **sweetly-scented flowers** appear in clusters in midwinter. The fruit is a small capsule.
LOGANACEAE Height: up to 3 m

Opuntia ficus-indica *
Sweet Prickly Pear

Large, leafless succulent with **sparse thorns on paddle-shaped stems**. A native of Mexico, this cactus was introduced to South Africa as a hedge and fodder plant but is now an invasive weed which despoils natural landscapes. It spreads via broken stems which take root, and by seed. Large, delicate flowers are pale yellow. Fruits are reddish and edible.
CACTACEAE Height: 1 to 2 m

Soft-stemmed Shrubs and Herbs

The Johannesburg area supports a remarkable variety of soft-stemmed plants but this section introduces only a few of the more common or interesting ones. Most of the photographs and descriptions emphasise the flowers of these plants, so they may not be easily identifiable at all times of the year. In addition to the indigenous flora, there are a great number of alien species which have become invasive weeds of disturbed soils. Since these species usually grow along roadsides and paths, they are often encountered, and are featured here for this reason. Further justification for their inclusion is that if they are recognised for what they are, some attempt may then be made to control their spread.

The identification of smaller plants may be a complex matter, and must often be undertaken in the most systematic manner. Since this is only an introductory guide, the traditional system of identification keys has been avoided in favour of a more simplistic approach which places plants with flowers of a similar colour, or of a similar form, together. As with trees, a useful learning technique is to **go out and look for the species featured**.

In some instances, plants which are represented by a number of similar species are identified to family level only (and are named: spp.); the accompanying photograph depicts a typical species. You will certainly come across plants not featured in this section, but knowing what they are not will help you identify them in a more detailed guide. The best publication dealing with the smaller plants of the area is the *Field Guide to the Wild Flowers of the Witwatersrand and Pretoria Region* by Braam van Wyk and Sasa Malan (Struik, 1988).

Aloe greatheadii (A. davyana)
Spotted Aloe

Stemless succulent with **fleshy leaves arranged in a rosette**. The leaves are spotted, have spiny margins and are often dry and twisted at their tips. The flowers vary in colour from salmon to red, and are produced on a spike in **winter**. Often form dense colonies in overgrazed areas. Resistant to fire. The similar *A. transvaalensis* flowers in summer.
ASPHODELACEAE

NG S

Aloe verrecunda
Grass Aloe

Erect succulent with **narrow fleshy leaves pointing upwards**. The leaves are finely spotted only on the underside, and the margins have small flexible 'teeth'. Grows among grass in rocky areas, and is easily overlooked when not in flower. The flowers are coral-red and hang on a short unbranched stalk during **summer**. The plant dies back in winter, sprouting anew with the onset of summer rain.
ASPHODELACEAE

KC NG

Striga elegans
Large Witchweed

Small upright herb with **dazzling scarlet flowers** produced in summer and autumn. A semi-parasite which derives part of its nourishment from the roots of grasses, including crops such as maize. The small slender leaves are held erect and are rough in texture. The closely related *S. asiatica* has much smaller, but equally bright, flowers.
SCROPHULARIACEAE

Zinnia peruviana *
Red Zinnia

Small annual herb, **native to South America**. A naturalised weed in places where soil has been disturbed, and also in the shade of bush clumps. Orange-red flowers are produced at the tip of single stalks during summer and autumn. The dry flower head may persist on the plant through winter. Slender leaves are in opposite pairs, spaced up the stem.
ASTERACEAE

Schizostylis coccinea
Scarlet River Lily

A beautiful bulbous plant with slender, strap-like leaves. The scarlet or pink, star-shaped flowers are produced on a drooping stem and are a lovely sight in semi-shade along streams or on damp cliffs in summer. Up to eight flowers occur on each stem. The plant dies back in winter. Uncommon, but conspicuous. Rewarding garden subject.
IRIDACEAE

Crossandra greenstockii
Orange Cross

Small upright herb with a short hairy stem growing from a woody rootstock. The flattened **orange flowers are clustered in a star-shaped head** of spiny overlapping bracts, and appear in spring or summer. The dark green leaves are oval, with wavy margins, and grow mostly in a rosette at ground level. Often grows in the shade of bush clumps.
ACANTHACEAE

Scadoxus puniceus
Royal Paintbrush Lily ☠

Beautiful bulbous plant with erect, **waxy, green leaves with wavy margins**. The leaves clasp the short, finely-spotted leaf stalk. The brilliant red, brush-like flower is a massed head of protruding stamens, and appears in spring. Green, pea-sized berries ripen to red. Prefers shade under trees. Fine garden plant but takes time to settle down and flower.
AMARYLLIDACEAE

NG RF

Haemanthus humilis
Rock Paintbrush Lily ☠

Beautiful bulbous plant with **flattened velvety leaves, with hairy margins**. The leaves appear at more or less the same time as the brush-like flower head – in midsummer. The flower may be pale pink or white and is smaller than that of the previous species. Green, pea-sized berries ripen to orange. Grows in shade among rocks.
AMARYLLIDACEAE

KC

Boophane disticha
Poison Bulb/Tumbleweed ☠

Bulbous plant with large, often fire-scarred bulb usually protruding above the ground. Pale green leaves are strap-like with wavy margins, and are arranged in a **distinctive fan pattern**. A massed head of crimson, star-shaped flowers appears in spring, before the leaves. Dry flower head is blown about by wind. Bulb contains poison.
AMARYLLIDACEAE

S NG

Crinum graminicola
Grass Lily

Bulbous plant with long, strap-like leaves spreading at ground level, and often trailing among grass. The spectacular funnel-shaped flowers are pink and white, and arise from a single stem in summer. The spherical fruits have a distinctive pointed tip. Occurs in open grassland. The related *C. bulbispermum* is restricted to damp or marshy areas.
AMARYLLIDACEAE

NG

Mirabilis jalapa *
Four o'clock Flower ☠

Waist-high shrub growing from a tuberous rootstock. Distinctive, **heart-shaped leaves** grow from four-sided stems. Beautiful, funnel-shaped flowers are clustered at tips of stems and may be pink, variegated or pale yellow. A **native of South America**, but now a naturalised weed of disturbed soils, particularly in **damp places**.
NYCTAGINACEAE

WR NG

Bidens formosa *
Cosmos

Feathery-leaved annual herb which bears daisy-like flowers in pink, maroon or white, with yellow centres. A **native of Central America**, but now a naturalised weed of disturbed soils, particularly along **roadsides and in old fields**. Grows in dense communities, and puts on a **magnificent massed flower display in autumn**.
ASTERACEAE

NG

Oxalis spp.
Pink Sorrel

Small perennial herbs with trifoliate, clover-like leaves. An underground rhizome may bear numerous small bulbs and the plant often forms dense colonies. Flowers are solitary, appearing on stalks longer than those of the leaves. At least three similar, pink-flowering species occur in the area. The yellow-flowered *O. corniculata* is an invader from Europe.
OXALIDACEAE

NG S G

Indigofera spp.
Pink Indigofera

Waist-high shrubs, shrublets or herbs usually with graceful pendant stems belonging to the pod-bearing, or legume, family. Feathery, compound leaves are widely spaced on stems. Pea-like flowers are pink, and borne in terminal heads in spring or summer. Pods are cylindrical or flattened. At least 15 species occur in the Johannesburg area.
FABACEAE

S NG KC

106

Ipomoea purpurea *
Morning Glory

Twining herb with showy, **funnel-shaped flowers** varying in colour from bright purple to white. A **native of Central and South America** but now an invasive weed of disturbed soils where it strangles other vegetation, or clambers up fences. Leaves are heart-shaped with pointed tips. Several indigenous species, with pink or pale yellow flowers, also occur in the area.

CONVOLVULACEAE

PS S RF

Verbena bonariensis *
Wild Verbena

Erect, sparsely-stemmed herb, **native to South America** but now a naturalised weed of disturbed soils along roadsides and often abundant in damp places. The stalkless leaves clasp stems. Small **purple flowers are in congested heads** at the end of branched stems during summer. The similar *V. brasiliensis* has stalked leaves and bi-coloured flower heads.

VERBENACEAE

NG WR

Verbena tenuisecta *
Fineleaved Ground Verbena

Carpet-growing herb, **native of South America** but now a naturalised weed of disturbed soils. Often **abundant along roadsides**, it puts on a showy flowering display during spring and summer. Bright purple or lilac flowers are arranged in terminal clusters; some may fade to white. Leaves are fern-like and hairy.

VERBENACEAE

S PS NG

Chironia palustris
Lilac Chironia

Beautiful, delicately-foliaged herb favouring damp or marshy soil where it grows in clumps up to waist-height. Star-shaped flowers are lilac or rose-pink with yellow centres, and are borne singly on multi-branched stems. The leaves are dark green and lance-shaped. The similar *C. purpurascens* has darker flowers, usually borne in groups of three, with a short-stalked, central flower.

GENTIANACEAE

WR

107

Xerophyta retinervis
Baboon's Tail/Firesticks

Stout perennial with erect stems often charred and blackened by fire. Rigid or arching, **grass-like leaves** grow in tufts from the stem in spring to form a willowy mound. The flowers are faintly-scented, funnel-shaped stars, varying in colour from lilac to white. Usually occurs in colonies in exposed positions on rocky ridges, mostly south-facing.
VELLOZIACEAE

Cyanotis speciosa
Doll's Powderpuff

Small, low-growing herb with long, boat-shaped leaves. All parts are covered in very fine hairs. The tiny but beautiful purple and white blooms are clustered in heads, with several to a stem, and have a woolly appearance and texture. Spring and summer is the flowering period. Usually grows in small colonies, often among moss, on south-facing rocky slopes.
COMMELINACEAE

Commelina spp.
Mouse Ears/Wandering Jew

Small, low-growing herbs with boat-shaped leaves – often hairy and widely spaced on trailing stems. Usually grows among grasses and other plants in shady places, but some species favour open rocky sites. The flowers are uniquely shaped with a pair of opposite petals. At least seven local species occur, with flowers being blue, mauve or yellow.
COMMELINACEAE

Moraea thomsonii
Lilac Tulp

Small, bulbous plant with **strap-like leaves**, occurring among grasses from which it is barely distinguishable when not in flower. Flowering stalk produces a single, **lilac, iris-like bloom** which opens in the afternoon but only lasts for a few hours before withering. Flowers in winter and spring. The closely related *Gynandriris simulans* has spotted flowers and grows only in marshy places.
IRIDACEAE

Pelargonium luridum
Stork's Bill Pelargonium

Small perennial herb usually growing among grass. Up to six leaves arise from the tuberous rootstock, and vary greatly in shape – plants may bear **undivided, toothed leaves**, or **deeply dissected leaves**. A single stalk produces a cluster of seven or more pendulous flowers in summer. The seed case resembles a miniature stork's bill.
GERANIACEAE

Tritonia nelsonii
Nelson's Tritonia

Small bulbous plant characterised by delicate orange flowers with distinctive, protruding outgrowths and a hooded, bonnet-like shape. Up to 12 flowers appear on each drooping stalk during summer. Usually found among grass in rocky areas. The **strap-like leaves** are ribbed, and arranged in a fan. Attractive garden subject.
IRIDACEAE

Gladiolus dalenii
Wild Gladiolus

Beautiful, erect bulbous plant up to one metre tall. The **sword-like leaves** are leathery with prominent ribs, and are arranged in a fan. The upright flower stalk bears a row of up to ten cone-shaped, orange or red **flowers in autumn**. Grows in small colonies. **Marvellous garden subject.** The related *G. crassifolius* bears pink or lilac flowers.
IRIDACEAE

Canna indica *
Canna Lily/Indian-Shot Canna

Broad-leaved lily with showy, but untidy, clusters of yellow, orange or red, trumpet-shaped flowers. Stems of up to one metre tall grow from an underground rhizome. The leaves are broad with a distinctive midrib. Widely-cultivated **native of tropical America** but now a garden escapee which grows along streams and in moist places.
CANNACEAE

Leonotis microphylla
Rock Wild Dagga

Low-growing herb with coarse, hairy leaves arranged in whorls. Orange tubular flowers arise from golf-ball-sized spikes, set at widely-spaced intervals on stems of up to two metres in height. Flowers appear in summer, but the dry spikes may persist into the next season. Grows among rocks. The related *L. dysophylla* is larger and grows in open grassland.
LAMIACEAE

Lantana camara *
Lantana/Tickberry

Tangled creeping shrub armed with prickles. Leaves are hairy and strongly scented. Small flowers are in flat-topped heads and are bi-coloured in orange and yellow or pink and yellow. Birds feed on the black berries and spread the seeds. A **native of tropical America** but now a very aggressive invasive weed which grows beneath and among trees.
VERBENACEAE

Rubus rigidus
Bramble

Sprawling tangled shrub armed with hooked prickles. Forms an impenetrable mass, often in disturbed soil or in bush clumps. Superficially similar to *Lantana*, but **leaves are trifoliate**, flowers are like miniature roses, and berries are raspberry-like. Several *Rubus* species from the northern hemisphere are serious pests in South Africa.
ROSACEAE

Clematis brachiata
Traveller's Joy

Sprawling vine which climbs vigorously with the aid of **tendrils** and may engulf supporting plants, or completely cover wire fences. Compound leaves have three or five leaflets, and are velvety in texture. The entire plant becomes massed with creamy-white flowers – consisting mostly of erect stamens – in **late summer and autumn**. Often in rocky places.
RANUNCULACEAE

Hypoxis spp.
Yellow Grass Stars
Perennial herbs which grow from an underground rhizome or corm. The star-shaped flowers are bright yellow, and are similar on all of the eight or more local species (which are known to hybridise freely). These plants are often the first to bloom on burnt ground in early spring. Leaves are strap-like and hairy, and may be erect or sprawling.
HYPOXIDACEAE

Hibiscus spp.
Wild Hibiscus
Perennial herbs or shrublets belonging to the cotton family and characterised by large flowers with thin, overlapping petals, numerous stamens forged into a tube and protruding anthers. Leaves of the eight or more local species vary in shape. Flowers are usually yellow, often with a maroon centre. May be confused with closely related *Pavonia* species.
MALVACEAE

Senecio spp.
Canary daisies
Perennial upright herbs with yellow daisy flowers at the ends of much-branched stems. The leaves of the over 20 local species vary greatly in shape but are often semi-succulent and blueish-green. Most species flower in spring or summer. Often appears as a garden weed. May be confused with several *Euryops* and *Osteospermum* species.
ASTERACEAE

Gazania krebsiana
Butter Star Daisy
Low perennial herb growing from a woody rootstock. Leaves are dark green, strap-like with a woolly underside and may exude a milky sap when broken. The radiant yellow (or white) daisy flowers appear in early spring, often on recently burnt ground. Each flower arises from a single stem. Could be confused with yellow-flowered *Gerbera* species.
ASTERACEAE

Helichrysum spp.
Everlasting daisies

Variable perennial daisies characterised by papery, **petal-like bracts which persist on the flower heads** after the seeds have been dispersed. Most species have yellow flowers, but plant form varies from erect to cushion-like. The leaves of some species are sticky, hairy or woolly. Over 30 *Helichrysum* species occur in the Johannesburg area.
ASTERACEAE

Dicoma zeyheri
Silver Everlasting

Small stout perennial with a woody root-stock. The stem is hairy and produces leathery, elongated leaves. The flower heads are produced at the tips of stalks and are surrounded by sharply-pointed bracts, which may persist for months. Appearing between midsummer and midwinter, the flowers resemble tiny silvery proteas tinged with pink.
ASTERACEAE

Bidens pilosa *
Blackjack

Upright, many-stemmed annual herb with opposite leaves consisting of three or five serrated leaflets. White and yellow daisy flowers are produced at the tips of stems. The distinctive **black seeds cling tenaciously to clothing**. A cosmopolitan weed, thought to be a **native of America**, which is a gregarious invader of roadsides, disturbed soils and gardens.
ASTERACEAE

Tagetes minuta *
Khakiweed/Khakibos

Erect annual herb with compound, **yellowish leaves** consisting of seven or nine leaflets. The small tubular flowers are set in upright heads, appearing in summer and autumn. The whole plant is **strongly scented**. A **native of South America**, but now an invasive weed of disturbed soils, gardens and roadsides where it is prone to form dense colonies.
ASTERACEAE

Cirsium vulgare *
Scottish Thistle
Robust and erect herb growing up to
waist-height. The **stems and jagged
leaves are armed with sharp spines**.
The purple flowers are produced at the
end of stems, and arise out of a prickly,
egg-shaped head. A **native of Europe**,
but now an invasive weed in many parts
of the world. Forms clumps in disturbed
soils, farmlands and rubble sites.
ASTERACEAE

Solanum sisymbrifolium *
Wild Tomato ☠
Much branched shrublet with **stems and
crinkly leaves armed with sharp spines**.
The flowers are lilac or white stars, with
back-facing petals and yellow centres.
Ripe fruits resemble tiny tomatoes but
are toxic. A **native of South America**,
but now an invasive weed of disturbed
soils, roadsides and rubble sites. Several
related species; some indigenous.
SOLANACEAE

Solanum seaforthianum *
Potato Creeper ☠
Trailing creeper which scrambles among
trees and bushes. The **compound leaves
are deeply dissected, with leaflets
often unevenly shaped**. The flowers are
small purple stars with yellow anthers.
Ripe fruits are small red berries, probably
toxic, which hang in bunches. A **native
of tropical America**, but now an invasive
weed.
SOLANACEAE

Datura stramonium *
Thorn Apple ☠
Robust, much-branched annual growing
up to waist-height. Leaves are large and
irregularly-shaped, and give off a foul
odour when crushed. Flowers are white
or pale mauve trumpets. Seeds held in
woody capsule, armed with stout spines. A
cosmopolitan weed, thought to be **native
to tropical America**, which grows in disturbed soils, farmlands and rubble sites.
SOLANACEAE

Asclepias fruticosa
Milkweed ☠

Slender perennial shrublet which may grow as a weed on disturbed soils and along roadsides. The **narrow leaves are lance-shaped** and occur sparsely on the branches. Small, creamy-white flowers hang in bunches during summer. Fruits are distinctive inflated sacs with sparse, rigid hairs. All parts exude a **milky latex**. There are at least 12 other local species.
ASCLEPIADACEAE

Xysmalobium undulatum
Bitterhout

Erect perennial herb with robust stems reaching shoulder-height. Occurs in open grassland, usually in the vicinity of water. The **stalkless leaves have wavy margins**, and a prominent pale midrib; both sides are roughly hairy in texture. **Waxy, cream-white flowers** are borne in heads in summer. Fruits are distinctive inflated sacs with sparse, curly hairs.
ASCLEPIADACEAE

Araujia sericifera *
Moth Catcher ☠

Sprawling vine which twines itself around the branches of trees or wire fences. The leaves have distinctive square bases. The attractive, star-shaped white flowers appear in summer. The fruit is a spongy green capsule which splits to release **silky seeds**; it dries to form a distinctive woody case. A **native of South America**, but now a naturalised weed.
ASCLEPIADACEAE

Stapelia gigantea
Giant Carrion Flower

Small, ground-hugging succulent with fleshy, finger-like stems. The leafless stems are green when the plant grows in the shade, but reddish and dry-tipped in full sun. The large, star-shaped flower is as wide as a man's outstretched hand, and emits a foul, carrion-like smell which attracts pollinating flies. A pair of long, double-horned capsules hold the seeds.
ASCLEPIADACEAE

Kalanchoe rotundifolia
Nentabos

Spindly succulent herb with fleshy leaves clustered at the base of the flowering stem. Usually grows in knee-high colonies in semi-shade at the base of trees and shrubs. Tiny urn-shaped, orange-red flowers are borne in dense, flat-topped clusters in autumn. The related *K. thyrsiflora* and *K. paniculata* are larger in size with yellow flowers.
CRASSULACEAE

Bryophyllum delagoense *
Madagascar Bells

Spindly, knee-high succulent with **spotted, grey-green, cylindrical leaves**. Spreads vegetatively with new plants growing from fallen leaves. Tubular, coral-red flowers hang in bunches.
A **native of Madagascar**, but widely cultivated and now an occasional garden escapee. Usually grows in colonies near habitation, and often in the shade of trees.
CRASSULACEAE

Polygonum spp.
Knotweed/Snakeroot

Slender annual herbs with a drooping form which flourish in damp places, often in water. Also grow in disturbed soils along roadsides. The leaves are lance-shaped, alternate and stalkless. **Pink or mauve flowers hang in drooping spikes**. The fruit is a small nut. At least five local species, one of which – *P. lapathifolium* – is an alien from Europe.
POLYGONACEAE

Phytolacca octandra *
Inkberry

Sparsely-branched herb or shrublet which grows in disturbed soils and along roadsides. Its origin is unclear although it is suspected to be **a native of America**. The fleshy **stems are red or purple** and support waxy, alternate leaves. Flowers are pale yellow and inconspicuous, but give rise to **spikes of fleshy, purple-black berries**.
PHYTOLACCACEAE

Lippia javanica
Fever Tea

Erect, much-branched perennial shrub often growing in colonies. The small coarse leaves are arranged in opposite pairs, and emit a **strong verbena scent when crushed**. Small white flowers are borne on long stalks in the leaf axils. The similar *L. rehmannii* rarely grows above knee-height and has lemon-scented leaves; it has short flower stalks.
VERBENACEAE

Protasparagus laricinus
Wild Asparagus

Scrambling, multi-stemmed shrub or climber with **flexible stems armed with spines** and often growing in a zig-zag manner. The leaves are needle-shaped but soft and springy to the touch. Tiny white flowers are clustered along stems in summer, giving off a strong sweet scent. Fruits are small, bright red berries. Five related species occur in the area.
ASPARAGACEAE

Trachyandra species
White Grass Stars

Slender, grass-like perennial herbs with an underground rhizome. Usually grows among tall grass, where its white, star-shaped flowers stand out in summer and autumn. At least four species occur in the region, but are easily confused with any of the five locally-occurring *Anthericum* species which are similar in appearance.
ASPHODELACEAE

Elephantorrhiza elephantina
Elephant's Root

Low-growing, **unbranched shrublet** with a woody, underground rootstock. The compound leaves are fern-like and green. The pale-yellow flowers are in dense spikes emerging at ground level in spring. Seeds are in **pods with persistent outer rims**. Forms colonies. The related *E. burkei* has blue-green leaves, is taller and flowers on the stem.
MIMOSACEAE

Scabiosa columbaria
Wild Scabious

Slender perennial herb with a woody
rootstock which grows among grass.
The stems and leaves are densely hairy.
Leaves are toothed or lobed and borne at
ground level. A single stem bears a
spherical head of numerous small white
flowers which attract nectar-feeding
butterflies. The closely related
Cephalaria zeyheriana is very similar.
DIPSACACEAE

Zaluzianskya katharinae
Pink Drumsticks

Small perennial herb with hairy, maple-
shaped leaves. The flower is unusual in
that it remains closed during the day,
when the club-shaped head resembles a
miniature drumstick. The **under-surface
of the petals is pink**. The white flowers,
with deeply-forked petals, open in the
evening. Grows among grasses, often in
rocky places.
SCROPHULARIACEAE

JO ONDERSTALL

Eulophia ovalis
Yellow Grass Orchid

Erect seasonal herb with a perennial
underground tuber. The leathery, strap-
like leaves appear on a shoot adjacent to
the flowering stem. Flowers open in
sequence on each stalk, and are white,
pale yellow or salmon with a flattened,
bonnet shape. Occurs in grassland, often
among rocks, with flowers appearing in
summer.
ORCHIDACEAE

Bonatea speciosa
Green Chandelier Orchid

Erect seasonal herb with a perennial
underground tuber. The fleshy, upward-
pointing leaves are arranged on the single stem. The spectacular flower head
carries a chandelier-style arrangement of
elaborate green and white flowers.
Blooms appear in late summer and
autumn. Usually grows in the shade of
trees and shrubs.
ORCHIDACEAE

Viscum rotundifolium
Wild Mistletoe

Fleshy parasite which attaches itself to the branches of trees. The evergreen stems are angular and jointed, the leaves rounded and leathery. Mature **plants are often ball-shaped**. The tiny flowers are inconspicuous, but the bright orange berries are prominent and relished by mousebirds. May flower and fruit at any time of the year.

VISCACEAE

Tapinanthus natalitius
Birdlime Bush

Parasite which attaches itself to the branches of *Acacia caffra* and other trees. Oval leaves are thick and leathery with soft grey hairs. The **upright, matchstick-shaped flowers are white** and appear in spring. The berries are green, ripening to black. The similar *T. rubromarginatus* is most common on *Protea caffra*, and has crimson-pink flowers.

LORANTHACEAE

Eichhornia crassipes *
Water Hyacinth

Free-floating aquatic plant with feathery roots. The **waxy, spherical leaves have swollen, bladder-like petioles**. Attractive lilac flowers are clustered on a spike. A native of tropical America but now an **invasive alien** which multiplies rapidly and may blanket dams and slow-moving streams. Poses a grave threat to aquatic habitats.

PONTEDERIACEAE

Myriophyllum aquaticum *
Parrot's Feathers

Mat-forming, rooted waterplant with long stems which send upright shoots above the water. The feathery leaves are arranged in whorls. The flowers are minute. A **native of tropical America**, but now an invasive weed which grows in shallow water. Only female plants are known to occur in South Africa, so reproduction is exclusively vegetative.

HALORAGACEAE

118

Typha capensis
Bulrush

Upright, aquatic herb which resembles a grass or a sedge but is not related to either. Usually grows in dense colonies alongside or in water. The leathery, strap-shaped leaves emerge from a creeping rhizome as do the tall, flowering stems. Flowers are densely packed on a cylindrical, velvety spike which splits to release masses of fluffy seeds.
TYPHACEAE

Grasses and Sedges

Grasses, of various kinds, are probably the most important of all plants, at least as far as mankind in concerned. All of our basic foods, including bread, maize, rice, cereals and sugar, are derived from grasses, and many foods and products are produced indirectly by grass grazers such as cattle and sheep. Apart from the two large species featured below, most grasses are difficult to distinguish from one another. Overleaf, eight of the more conspicuous 'typical' grasses are illustrated, but no attempt is made to discuss their characteristics. An excellent book entitled *Guide to Grasses in South Africa* by Frits van Oudtshoorn and Eben van Wyk (Briza, 1992) is valuable reading for anyone wishing to make a more detailed study of these important plants.

Phragmites australis
Common Reed

Tall perennial grass with creeping rhizomes which grows in permanently waterlogged situations in wetlands, usually in **dense colonies**. Thin papery leaves are sheath-like and have razor-sharp margins. The white flowers are produced in massed spikes at the tips of the stems. These plants are known to 'purify' water by absorbing pollutants.
POACEAE

Cortaderia selloana *
Pampas Grass

Tall perennial grass with strap-like, razor-sharp leaves which **grows in small clumps**. The flower heads are white and set in an elegant spike. A **native of South America**, it is grown as an ornamental in gardens, but is now a naturalised escapee of disturbed soils, often near water. The related *C. jubata** has pinkish flowers and is cultivated on mine dumps.
POACEAE

***Themeda triandra* Rooigras**
Tufted grass with triangular flower heads
becoming red. Important grazing species.

***Melinis repens* Redtop Grass**
Tufted grass with delicate pink or red
flower heads. Common along roadsides.

***Hyparrhenia anamesa* Thatch Grass**
Tall tufted grass with spiky flower heads
on stiff stem. Common along roadsides.

***Heteropogon contortus* Spear Grass**
Tufted grass with spear-like flower heads
which often become entangled.

***Elionurus muticus* Wire Grass**
Tufted grass with flowers in a woolly spike
which curls over when ripe. Unpalatable.

***Pennisetum clandestinum* * Kikuyu**
Creeping grass native to East Africa. Used
widely for lawns and pasture, but invasive.

***Mariscus congestus* Giant Sedge**
Large, robust perennial which grows up
to 2 m on wetland fringes.

***Cyperus rupestris* Dwarf Sedge**
Small, tufted perennial which grows in
shallow soil, often among rocks.

Ferns and Primitive Plants

Ferns are perennial plants usually growing from an underground rhizome, and most species prefer damp, shady places. Lacking flowers, reproduction is by tiny spores contained in brown-coloured structures called sori which are arranged in rows under the surface of the fronds (leaves). Fronds of most species are soft in texture and deeply divided.

Over 30 species of fern occur in the Johannesburg area; the four illustrated below being fairly common. The most authoritative publication on these interesting plants is *Ferns of Southern Africa* by John Burrows (Frandsen, 1990).

Many species of moss and lichen occur in the area, but their identification is usually a technical matter beyond the scope of this book.

JOHN BURROWS

***Pellaea calomelanos* Hard Fern**
Adapted to exposed conditions. Rigid, blue-green fronds.

***Mohria caffrorum* Parsley Fern**
Grows in sheltered spots beneath rocks or shrubs. Soft, hairy leaves.

***Salvinia molesta* * Kariba Weed**
Invasive alien from tropical America. Multiplies rapidly, covering water surface.

***Pteridium aquilinum* Bracken**
Large, coarse-leaved fern. Cosmopolitan. Forms colonies on slopes or damp sites.

Mosses

Mosses are tiny green plants, which grow in furry-textured 'mats'. Able to withstand dry periods, but typical of moist places.

Lichens

Lichens are so-called 'dual plants' which consist of a fungus growing in association with an alga. Cling to rocks or bark.

REFERENCES AND FURTHER READING

GENERAL

Carruthers, V.C. (ed.) 1982. *The Sandton Field Book.* Sandton Nature Conservation Society, Rivonia.

Ryan, B. 1987. *52 Day Walks in and around Johannesburg and Pretoria.* Struik, Cape Town.

GEOLOGY

Mendelsohn, F. & Potgieter, C.T. 1986. *Guidebook to Sites of Geological and Mining Interest on the Central Witwatersrand.* Geological Society of South Africa, Johannesburg.

Pritchard, J.M. 1986. *Landscape and Landform in Africa.* Edward Arnold, London.

MAMMALS

Apps, P. 1992. *Wild Ways: Field Guide to the Behaviour of Southern African Mammals.* Southern Books, Halfway House.

Stuart, C. & T. 1988. *Field Guide to the Mammals of Southern Africa.* Struik, Cape Town.

Smithers, R.H.N. & Abbott, C. 1992. *Land Mammals of Southern Africa: A Field Guide.* (2nd edition) Southern Books, Halfway House.

Skinner, J.D. & Smithers, R.H.N. 1990. *The Mammals of the Southern African Subregion.* (2nd edition) University of Pretoria, Pretoria.

BIRDS

Chittenden, H. 1992. *Top Birding Spots of Southern Africa.* Southern Books, Halfway House.

Gibbon, G. 1991. *Southern African Bird Sounds* (set of 6 cassettes). Southern African Birding cc, Durban.

Ginn, P.J., McIlleron, W.G. & Milstein, P. le S. 1989. *The Complete Book of Southern African Birds.* Struik Winchester, Cape Town.

MacLean, G.L. (ed.) 1993. *Roberts' Birds of Southern Africa* (6th edition) John Voelcker Bird Book Fund, Cape Town.

Newman, K. 1992. *Birds of Southern Africa* (4th edition). Southern Books, Halfway House.

Ryan, B. & Isom, J. 1990. *Go Birding in the Transvaal.* Struik, Cape Town.

Sinclair, I., Hayman, P. & Arlott, N. 1993. *Sasol Birds of Southern Africa.* Struik, Cape Town.

Tarboton, W.R., Kemp, M.I. & Kemp, A.C. 1987. *Birds of the Transvaal.* Transvaal Museum, Pretoria.

Trendler, R. & Hes, L. 1994. *Attracting Birds to your Garden in Southern Africa.* Struik, Cape Town.

REPTILES

Branch, B. 1988. *Field Guide to the Snakes and Other Reptiles of Southern Africa.* Struik, Cape Town.

Broadley, D.G. 1983. *Fitzsimons' Snakes of Southern Africa.* Delta Books, Johannesburg.

Patterson, R. & Bannister, A. 1987. *South African Reptile Life.* Struik, Cape Town.

FROGS

Carruthers, V.C. 1976. *A Guide to the Identification of Frogs of the Witwatersrand.* Conservation Press, Johannesburg.

Passmore, N.I. & Carruthers, V.C. 1995. *South African Frogs* (2nd edition). Southern Books, Halfway House &Wits. Univ. Press, Johannesburg.

Wager, V.A. 1986. *Frogs of South Africa: Their Fascinating Life Stories.* Delta Books, Craighall.

FRESHWATER FISHES

Skelton, P.H. 1993. *A Complete Guide to the Freshwater Fishes of Southern Africa.* Southern Books, Halfway House.

INVERTEBRATES

Pringle, E.L.L., Henning, G.A. & Ball, J.B. (eds) 1994. *Pennington's Butterflies of Southern Africa.* (2nd edition) Struik, Cape Town.

Filmer, M. 1991. *Southern African Spiders: An Identification Guide.* Struik, Cape Town.

Holm, E. & de Meillon, E. 1986. *Pocket Guide: Insects.* Struik, Cape Town.

Migdoll, I. 1987. *Field Guide: Butterflies of Southern Africa.* Struik, Cape Town.

Newlands, G. & de Meillon, E. 1986. *Pocket Guide: Spiders.* Struik, Cape Town.

Pinhey, E.C.G. 1975. *Moths of Southern Africa.* Tafelberg, Cape Town.

Scholtz, C.H. & Holm, E. (eds) 1985. *Insects of Southern Africa.* Butterworths, Durban.

Skaife, S.H. (revised by Ledger, J.A.) 1979. *African Insect Life.* Struik, Cape Town.

Williams, M. 1994. *Butterflies of Southern Africa: A Field Guide.* Southern Books, Halfway House.

PLANTS

Burrows, J.E. 1990. *Ferns of Southern Africa.* Frandsen, Johannesburg.

Coates Palgrave, K. 1983. *Trees of Southern Africa.* Struik, Cape Town.

Fabian, A. & Germishuizen, G. 1982. *Transvaal Wild Flowers.* Macmillan, Johannesburg.

Johnson, D. & S. 1993. *Gardening with Indigenous Trees and Shrubs.* Southern Books, Halfway House.

Henderson, M. & Fourie, D.M.C. 1987. *Declared Weeds and Alien Invader Plants in South Africa.* Dept of Agriculture and Water Supply, Pretoria.

Lucas, A. & Pike, B. 1971. *Wild Flowers of the Witwatersrand.* Purnell, Cape Town.

Tree Society of South Africa. 1974. *Trees and Shrubs of the Witwatersrand.* Wits. Univ. Press, Johannesburg.

Van Gogh, J. & Anderson, J. 1988. *Trees and Shrubs of the Witwatersrand, Magaliesberg and Pilanesberg.* Struik, Cape Town.

Van Oudtshoorn, F. & van Wyk, E. 1992. *Guide to Grasses of South Africa.* Briza, Pretoria.

Van Wyk, B. & Malan, S. 1988. *Field Guide to the Wildflowers of the Witwatersrand and Pretoria Region.* Struik, Cape Town.

USEFUL CONTACT ADDRESSES

CONSERVATION

Directorate of Culture and Recreation: Conservation Department
c/o Johannesburg Zoo, Jan Smuts Ave, Parkwood. Tel. (011) 646 2000
Johannesburg Parks Dept.
Tel. (011) 407 6801
Sandton Parks, Recreation and Conservation
PO Box 78001, Sandton 2146.
Tel. (011) 803 7917/8 or 803 9300/1
Wildlife Society of SA
PO Box 44189, Linden 2104.
Tel. (011) 486 0938/9
Endangered Wildlife Trust
P/Bag X11, Parkview 2122.
Tel. (011) 486 1102

EDUCATIONAL

Delta Environmental Centre
P/Bag X6, Parkview 2122.
Tel. (011) 888 4831
Johannesburg Zoo
Jan Smuts Ave, Parkwood.
Tel. (011) 646 2000
Transvaal Museum
PO Box 413, Pretoria 0001.
Tel. (012) 322 7632
Sappi-Brett Enviro. Courses
PO Box 650727, Benmore 2010.
Tel. (011) 783 6629

BIRDS and BIRDWATCHING

SA Ornithological Society
PO Box 87234 Houghton 2041.
Tel. (011) 888 4147
Witwatersrand Bird Club
PO Box 72091, Parkview 2122.
Rand Barbets Bird Club
135 Oxford Road, Saxonwold 2193.
Sandton Bird Club
PO Box 650890, Benmore 2010.

REPTILES

Transvaal Snake Park
PO Box 97, Halfway House 1685.
Herpetological Ass. of Africa
PO Box 266, Bloemfontein 9300.

SPIDERS

Spider Club of South Africa
PO Box 51750, Raedene 2124.

BUTTERFLIES

Lepidopterist's Society of SA
PO Box 470, Florida Hills 1716.

INJURED ANIMALS

Animal Rehabilitation Centre
PO Box 15121, Lynn East 0030.
Tel. (012) 808 1106
Johannesburg Zoo
Jan Smuts Avenue, Parkwood.
Tel. (011) 646 2000

PLANTS

Botanical Society of SA
Kirstenbosch, Claremont 7735.
Tel. (021) 797 2090
    JHB Branch Tel. (011) 788 7541
    Bankenveld Tel. (011) 678 6890
Dendrological Foundation
PO Box 104, Pretoria 0001.
Tel. (012) 57 4007
National Botanical Institute
P/Bag X101, Pretoria 0001.
Tel. (012) 804 3200
Trees for Africa
PO Box 447, Strathaven 2031.
Tel. (011) 337 3000
Witwatersrand Bot. Garden
PO Box 2194, Wilro Park 1731.
Tel. (011) 662 1741
Witkoppen Nursery
PO Box 67036, Bryanston 2021.
Tel. (011) 705 2703
Buffelskloof Forest Nursery
PO Box 710, Lydenburg 1120.

FRESHWATER FISHES

J.L.B. Smith Institute of Ichthyology
P/Bag 1015, Grahamstown, 6140.
Blyde River Aquaculture
PO Box 408, Hoedspruit 1380.
Tel. (016) 63131
(suppliers of indigenous fishes)

GLOSSARY OF SCIENTIFIC TERMS

alien – an organism introduced by man, and now naturalised in a region or country in which it does not belong
aquatic – living in water
arboreal – living in trees
alternate – leaves which are arranged singly at different points on a stem
annual – a plant which completes its life cycle within one year
anther – pollen-bearing part of a flower
axil – upper joint between a leaf and a stem
biennial – a plant which grows and develops in the first year, and sets seed in the second
bipinnate – a compound leaf in which the leaflets are further divided into pinna (eg *Acacia*)
bract – leaf-like structure from which a flower arises
compound – a leaf consisting of several leaflets (eg *Cussonia*)
crepuscular – active at twilight, or just before dawn
deciduous – a plant which sheds leaves at the end of the growing season

dorsal – upper surface of the body
dorsal fin – fin on the spine of a fish
drupe – a fleshy, non-splitting fruit
epiphyte – an organism that grows on another but is not parasitic
food chain – the sequence whereby plants are consumed by herbivorous animals which are then preyed upon by other animals
gills – breathing organs of fishes
herbivorous – eating plant matter
indigenous – an organism occurring naturally in an area
latex – a white, sticky liquid
leaflet – divided leaf
mimic – one animal resembling the form or colour of another, in order to obtain some benefit
naturalised – an organism which has been introduced from elsewhere and is reproducing successfully in a new area
native – *see* indigenous
opposite – leaves which are arranged opposite to one another on a stem
parasite – an organism which obtains its food from another organism (host)

perennial plant – a plant which lives for at least three years
perennial river – a river which flows throughout the year
pinna – divided part of a leaflet
pinnate – a compound leaf divided into leaflets
rhizome – a creeping, underground stem
roost - night-time rest place of birds or bats
simple leaf – an undivided leaf
scale – a thin, plate-like structure
scalloped – leaf margin notched with blunt projections
spike – an elongated stem which bears more than one flower
serrated – leaf margin notched with fine projections
terminal – at the end of a stem
terrestrial – living on the ground
toothed – leaf margin notched with pointed projections
trifoliate – a leaf which is divided into three leaflets (eg *Rhus*)
ventral – under-surface
whorled – the arrangement of three or more leaves or flowers at the same point on a stem to form an encircling ring

INDEX OF FEATURED SPECIES

125

126

Winter Into Spring

I dream I'm trying to unlock the door to my car and get inside—should that be so difficult? But there are children in the parking lot or on the street where I am trying to accomplish this—little boys—and they are running wild and out of control, getting in my way as I try to get into my car.

"Move your monkeys!" I shout at their mother, who is looking on. "Keep your monkeys under control!"

One of the little boys looks at me jeeringly. "Monkeys are always girls, and I'm a boy, so I can't be a monkey."

"Of course monkeys can be boys, you little fuck," I say.

"Mommy, that woman said 'fuck.' "

I turn to the entire family now gathered round and gaping at me and yell, "Fuck! Fuck! Fuck! Fuck! Fuck!"

And then the weather changes. It's been a long, slow crawl, but spring is finally here. Each year it seems entirely possible that winter will never ever leave, and why should it? It only has to return four months later. Why even bother to

pack its bags and head south of the equator. But then a crocus pops its yellow or blue or purple or pink head up from the earth, to signal the daffodils that the coast is clear. The daffies alert the forsythia and they send up a golden "GO!" to the magnolias and the dogwoods and soon there are leaves on the trees and it's hard to remember why there's a shovel and a bag of ice-melt on your porch.

Faith and I walk whenever we can. We spend a weekend in Manhattan wandering around the lower east side and Chinatown, in and out of department stores on Broadway. We crisscross the working-class blocks of our neighborhood, tour around the Victorian-lined streets near my mother's town house. We walk up and down hills at the arboretum, weaving our way around cherry trees and white spruce, hemlock trees and black walnut.

There are a hundred million dogs in the arboretum, each being led by a lesbian couple and their child. It's true. Fo every flowering blossom there is a lesbian, and for every le bian there is a baby. It's overwhelming, mind-boggling, epidemic. If one were only to frequent parks like the Arr Arboretum one might surmise that all American fan consisted of two moms in jeans, a baby in overalls, Labrador-shepherd mix. Just as there are among hete ual families, there are kind-looking lesbian families, ried and overworked, the gentle and relaxed, and and aloof. We are just another couple among m children, should we ever have them, will be just among kids. So long as we don't venture too f Arnold Arboretum, all will be well. We will be just like the scenery.

makes me feel really, really guilty, like I drove him away, like I was such a bitch he was afraid even to ask if he could stay with us. I'm hurt and relieved, but mostly I worry Eric has decided we are strange lesbians, communal-woman and hermit-girl, trying to make a life together. That must be why instead of being around the freakiness of our union he has chosen to stay with a straight couple and their two children.

Faith and I are driving past the very spot where last fall we had to pull over so we could have our freezing-cold fight, where Faith had to blow her nose in a sock and I had to pretend it didn't gross me out because we were talking about heavy things like whether she really wanted a child. Now she tells me that over lunch today, Eric—not on American soil for more than twelve hours—announced he and his girlfriend had spent many long Spanish hours together talking and have decided they want to offer us Eric's sperm, if we're interested of course, because they love us and think we're a great couple and would really like to help us to have a baby.

I'm so relieved that there are no hard feelings, that Eric took to heart my taking him aside last December and telling him how sorry I was about my moodiness and that I loved him and that I just go crazy when someone stays with us for longer than five days. I'm so relieved that I hardly hear what he's offered. After all, there are two vials of five-foot ten-inch half-Japanese sperm and a week's worth of ultrasounds waiting for me at the new fertility clinic. After all, we have decided on an anonymous donor. We've been through this known-donor thing so many times already, and we are done, done, done with it.

Or are we?

Faith is trying not to convince me of anything, trying not to push, trying to communicate Eric's offer without conveying how very much she wants to accept it. But I know. I know her, and I know how much she cares for this man, how much she wishes we could know our child's father.

I am thinking this: known donor versus cancer. Which would I prefer? What if all I need to get pregnant is fresh sperm, a week's worth of the good stuff swimming around in me? What if it's the single-shot sniper fire of frozen sperm and the sheer medical stress of going into an exam room and donning a paper skirt that has my reproductive system all twisted and out of shape? What if we just load me up this week with fresh sperm and see what happens; if I get pregnant, it was meant to be. If I don't, it's back to the Clomid drawing board. I mean, how better to determine the state of my fertility than to tank up on stuff that lives for five days and looks perky and crisp like a fresh Florida grapefruit?

Come on, everything happens for a reason.

Besides, aren't I grateful for Faith's communal nature? Hasn't it saved me from a life of excessive quiet and televisionless solitude? Hasn't it brought me new experiences? Hasn't it challenged me? Hasn't it introduced us to an assortment of artists and musicians and actors who fill our life with creativity and adventure? Hasn't it brought me rock operas and children's theater and Sunday evenings with *The Sopranos*?

After dinner I take a long walk around our neighborhood. It seems to me on this warm May evening that what happened with Eric when he was in last time could have been avoided if I had dealt with him directly, if I had confronted him with my feelings and not tried to use Faith as an intermediary, rather than trying to get her to feel exactly what I was feeling exactly when I was feeling it. If I can talk to Eric about that, then I will feel more in control of my boundaries and maybe even able to imagine a lifetime of negotiation and communication with him. A lifetime...an entire lifetime, of communicating and sharing a child. With Eric?

But anyway, the frozen-sperm thing *is* getting pretty old. The doctors and the ultrasounds and the ovulation test kits, the nine other lesbians who will be raising our children's half-siblings, the money, the money, money, money, it's all old and stale and stressful. What if our children could know their father? What if it could be as easy for Faith and me to get pregnant as it was for our sisters and our mothers? What if we didn't even go into months' worth of discussion and just did it? What if we called Eric over there at the heterosexual house where he is staying and said, "All right. Come over tomorrow and we'll try. We'll try every day for a week." What if we leapt before we looked? In one month I will turn forty. If ever there was a time for leaping, it's now. And I would so much rather leap into the arms of a five-foot six-inch Jewish friend than a bottle of potentially cancer-causing Clomid.

While we're at it why not shoot up with all five of Baldie's remaining vials? I'm not sure what opposing spermatozoa do when they run into each other in a woman's uterus or fallopian tube. I imagine it is not too unlike an Olympic Freestyle competition or the Indianapolis 500. Maybe the competition even would spur Baldie on. Sometimes men like to feel jealous. It's an evolutionary Darwin thing.

Simone—with the help of Tory—calls me at work to propose. Our wedding, apparently, will be a private event involving me in a "twirling dancing skirt" and Simone in a princess costume. There will be a party—for just the two of us—and then the wedding. She is not yet four, but I'm certain we were meant to be together. That, and if my own child evoked in me as much—or more!—love as does this small friend, then it's all over. I would explode or implode or drown in a sea of heart-shaped tears.

I'm not sure about Eric. I remember the expression on his face when he learned we had only sandwich-size baggies and not quart-size, the exasperation and the disdain.

And even if I become the best communicator in the world, what will it mean to constantly have to set limits for and demand boundaries of the father of our child?

Maybe Faith and Eric are too close for this to be comfortable. Didn't we agree that the right man was a cold, arrogant stud?

Wait a minute. Didn't we agree we were done with this decision?

After five days of blood tests I'm developing tracks on both of my arms. Repeated ultrasounds are like having a window installed in my womb, like being a reproductive Teletubby. The new and improved fertility clinic charts my hormonal comings and goings with the precision of a paramilitary task force. They are all about not shooting until they see the whites of my egg. They are all about getting me pregnant. It's them I trust, them I love. It's the new fertility clinic who will be the father of our child.

Eric has been at the house each evening this week for an hour here, a half-hour there, certainly long enough to produce a little genetic material. Faith has not said another thing about it. Nor have I. Nor has Eric. Now that the initial haze of gratitude and fantasy has worn off perhaps I am not the only one with reservations. I drop the subject, and much to my relief, nobody decides to pick it up again.

My mother calls to tell me she has scheduled a blood transfusion. I wish there were some way to give her the blood left over from my many tests. It's so bizarre a symmetry, my giving blood, her getting blood. Anyway, the chemo has depleted her red count, and her liver is just too traumatized to assist in the manufacture of new cells, so she is pooped. A blood transfusion is not necessary, her doctor explained, but it might give her more energy. So, "sure," she said. "Why not?" Six hours later she feels great.

"I wonder whose blood it was," my mother muses with her new energy. "It's strange to have something inside you that once was inside someone you don't even know."

I can't help consider how grateful and reliant we each have become on the bodily fluids of strangers. Blood is so much bigger, so much deeper and redder than semen, but you don't have to pore through donor catalogs before you have some infused into you.

"I know what you mean," I tell her.

We are infested with carpenter ants. They are *everywhere*: in the pantry, the cat's bowl of food, the dishwasher. I find one in the bathroom heading toward the tub, and a couple more walking from one sofa to another in the living room. I metamorphosize from a touchy-feely earthy-crunchy who carefully slips moths and spiders into Dixie cups and takes them back outside to live among their loved ones, into a killing machine. I kill three ants at a time. I kill them while they're eating. I kill them while they're communicating meaningful ant information to one another. I kill them while they're trying to run away from me. Faith and I kill and kill and kill until one day she calls me at work and announces, "I can't take it anymore," and then I call an exterminator.

The fact that I will consent to having ant-killing carcinogens oozed and sprayed and blown into our house but will not take Clomid or other fertility enhancing drugs is not lost on me. Not an hour goes by when I do not question whether or not I am more interested in killing ants than having a baby. The exterminators assure us that after four hours the poison is no longer harmful to humans. Four hours after every ant in our house has been chemically slain in the ant equivalent of a nuclear apocalypse, our house will revert back to the organic produce section at Whole Foods. I open all the windows, take all of our dry goods and spices, bring the kitty to Tory's house, and hope for the best.

That evening, with the exception of a valiant few who lie mangled and dying, all of the ants are dead. I feel sad, guilty. I imagine smoldering ruins, little ant war correspondents taking pictures of a blitzkrieg that will go down in the annals of ant history, that will be told by crusty old ants to their grandchildren sixty years from now in somebody else's kitchen. "It was a sunny day in May 2001, and not one of the fourteen million of us living in that house saw it coming." I say a prayer and ask for forgiveness—they are living creatures, after all.

At 2 P.M. a nurse calls me at work to announce I am surging. "Guess what, Harlyn?!"

In the world of trying-to-conceive news, your LH surge is first runner-up to the crowning moment when you find out you're pregnant. It's the next best thing, the prerequisite. It's renewed hope and the beginning of two more weeks during which you actually might be pregnant, as opposed to those two weeks when you know for a fact that you are not. It's optimism and control and a sign that there's every reason in the world to believe your body is in full working order and that this time will be the one.

The nurse schedules me for two consecutive IUI's—I don't even have to plead for the second one. At the new and improved fertility clinic it's two or none. They refuse to do just one.

I prepare to cast away the odious habits I have fallen into since the last time we tried: two cups of caffeinated tea a day, staying up until midnight to watch back-to-back episodes of *All in the Family.* I'm on the fertility wagon as of this very moment. It's clean living for the next fourteen days.

"Good luck, honey," the nurse says.

At the new and improved fertility clinic they call you honey.

The speculum does not break inside me. The nurse does not jab my uterus with a catheter and then roll her eyes and claim, "I hit a brick wall." Even the thawed semen is a pretty shade of pink instead of the creamy yolky yellow that it was at the previous clinic. In addition to calling me honey, the new fertility clinic even insists I lie still and relax for fifteen minutes after the insemination.

It's the first time *ever* that I've had non-Jewish sperm inside me. I imagine a school of uncircumcised spermatozoa crossing themselves before swimming toward my little Jewish egg. I hope they're not anti-Semitic, these Catholic-Buddhist sperm. I hope they treat my egg with respect and roll back their foreskins before doing the deed. But I know they will make it. They are not bald. They are not so similar as to be invisible. They are not loaded with Tay-Sachs and other Ashkenazi anxieties. They are mixed-race, potent, horny cross-breeds who run downstairs on Christmas morning all piss and vinegar and know the proper way to hold a pair of chopsticks. They're going to do it, I know they will. Because neurotic thoughts of not doing it will never even cross their tiny non-Jewish XY minds.

I catch myself feeling optimistic and hopeful, but because I *am* Jewish, I immediately slap my own hand and suppress all positive thoughts. *Kenahora.* There is even a Yiddish term for what I am doing. I am trying not to give myself a *kenahora,* i.e., jinxing myself with good thoughts. Of course, whether or not I get pregnant will have nothing to do with whether or not I have wished for the best. In fact, even if my thoughts and my fertility were related, then feeling hopeful and optimistic would do much more to encourage my body to become a warm receptive, place than would feeling numb and negative. I try to tell myself to let go, to feel as much hope and optimism as I want. I'm not in control, and it's not like I can ward off disappointment by tricking myself into expecting the worst.

Perhaps I'm afraid that if I really let myself know how much I want this then who knows what more I will do to achieve it—maybe take Clomid, maybe give myself HCG injections or have my eggs harvested. Maybe I will endanger myself. Maybe I will go too far. Or worst of all, I might join the masses of women who *want* a baby, and that is just shades away from the abyss of women who power-walk in malls and wear long, flowing shirts over spandex stretch pants.

After fifteen minutes I roll off the table and walk gingerly into the bathroom to change.

IUI number two of the first insemination of the rest of my life is performed by the same kind and gentle nurse-midwife of the previous day. She shows me our donor's number, the name of the sperm bank, his counts.

"His motility is not as high as we'd like," she says, just in case I had started to allow myself some semblance of optimism. "It was a little low yesterday, and today it's even lower."

Lying there half-naked in a johnny after a week's worth of blood tests and ultrasounds, with my fragile optimism balled up in a sweaty little wad at the back of my throat, it's hard not to feel that maybe I'm just not supposed to get pregnant. And maybe that doesn't have anything to do with being gay; other gay women get pregnant. No, maybe it's personal. Maybe it's just something to do with me and the universe, some quirk of reincarnation or karma. Either way, it makes me so sad.

The nurse-midwife gently guides my feet into the stirrups, because it's worth a shot, apparently. Even twenty-five percent of 248 million is something—only one little sperm has to have enough get up and go to swim from my uterus into my fallopian tube, and there have the wherewithal to penetrate my egg membrane. Only one. The rest can lie around like the dying, disfigured ants that have been limping around our kitchen counters ever since the massacre.

I position one foot and then the other inside the friendly, feminist oven mitts that are draped over the cold steel patriarchal stirrups and lean back. The nurse-midwife seems a million miles away. It's both intimate and desolate, like oral sex: The person with whom you are sharing one of life's most private moments is south of your equator in a whole other time zone, while you are way up north like a reindeer. It's lonely like that now with the midwife between my legs busily inserting a speculum. I feel so small and alone and incapable of making the choices required to create a family. My girlfriend is at work, my mother has cancer, and the second of two very expensive and time-consuming donors may be a bust.

"I'm threading the catheter now."

"Okay."

"I'm injecting the sample. How are you doing?"

I want to say that I'm exhausted and kind of bummed out. It's all just too much, this frozen-sperm thing. Do you think thirty-nine-year-old lesbians should even be trying to get pregnant? And how about the infertile heterosexual couples you see—what about them? How far do you think women should go? Is it good that we can postpone pregnancy until we're in our tired forties, at which point we can decide to have eggs harvested, fertilized in a petri dish, and transplanted back into our wombs, or maybe even the womb of a stranger? Are multiple births, old parents, and the dwindling of that wonderful phenomenon known as grandparents a good thing?

"I'm doing fine," I tell her. I'm glad she's so far away, because I don't want her to see the tears welling up in my eyes. It's enough to know that I want something I may not get and another thing entirely to begin confessing that to strangers, even very kind and gentle midwives who have heard it all before.

I'm not angry this time. I'm not considering suing or asking for our money back. I just think they should know, the sperm bank, that is, about these low counts, so that what happened the last time with Baldie does not happen again. I'm convinced now it's the nature of the game—frozen sperm cannot compete with the fresh stuff. Even otherwise potent, fertile sperm cannot always stand up to being frozen for six months, thawed 3,000 miles away from home, and then bathed in a chemical wash. It's a lot for a gal to ask of a guy.

The woman at the new sperm bank (do men ever work at sperm banks?) offers me a lesson on the nuances of frozen sperm. She explains that motility is less of an issue with frozen sperm than with fresh, that motility doesn't reflect the sperm's forward motion. It reflects rather the number of live sperm. And because something or other already has been done or assessed in the process of freezing semen, that number is a portion of something bigger than it would be if the sperm were fresh. I have no idea what she is talking about. It doesn't really make sense, in fact seems counterintuitive and in direct opposition to what the nurse-midwife told me, but I'm relieved nonetheless. I'm relieved because the motility—while low according to the fertility clinic—is actually standard for the sperm bank. It's what they expect from frozen semen.

"Besides," says the woman. "We had two reported pregnancies with this donor in just the last two months."

Well, now! Score! Baldie with his incredible counts sired no one. In this game there's only one thing that counts and that's pregnancy. The donors you want are the ones who have gotten somebody pregnant. It's as simple as that. What good are a zillion live and mobile sperm if they're defective in some way, lazy, tired, mutant? What good is a sharp-shooter if he's shooting blanks?

I walk gingerly the rest of the day so as not to jiggle those twenty-five percent of 248 million live wires out of my body. When the reproductive endocrinologist overseeing our treatment calls the next day to warn that under the circumstances (my age, the frozen sperm) there is only a five percent chance that I will become pregnant this round, I hardly care. Five percent. Somebody must fall into that five percent. Well, why not me and my frozen fertile boyfriend? *Kenahora.*

My sister calls to tell me about her amnio. She describes it as the strangest feeling she ever has experienced, like having someone press down really hard on her belly button, only deeper. She says she did not look at the needle and asked over and over again, with her face turned into the exam table, "Does everything look okay? Does everything look okay?"

Everything looked fine. The final results will be in next week.

"Piece of cake," she says. "You'll do fine."

"I'm not even pregnant."

"You will be."

If, statistically speaking, there is only a five percent chance that I'm pregnant, there's still only a twenty percent chance that a healthy heterosexual woman copulating with her fresh sperm spouse each night will conceive. It's incredible anyone gets pregnant. Maybe it's nature's way of enforcing population control. Otherwise there would be zillions of us, humans would be covering the earth like kudzu. In the middle of these

---

staggeringly disappointing odds is Clomid—the potentially cancer-causing candy that catapults a woman's chance of becoming pregnant with frozen sperm to a practically "normal" fifteen percent. The trick to getting pregnant, therefore, seems to be to keep trying. Trying, trying, trying, trying. The urban myth that women who have not been able to get pregnant suddenly will conceive after they adopt a child might simply be a reflection of rule number one: It takes a long time to get pregnant, even in the best of circumstances.

To distract myself from obsessive thoughts over the statistics of fertility and human population growth as well as my body's various minute aches and pains and secretions, I decide to spend my lunch hour at Starbucks. A chai latte seems safely removed from sperm and catheters and hormone levels, until a young mother and her small son sit at the table next to me and have this conversation:

"Mommy, tell me a story."

"About your birth?"

"Yeah."

"Well, when you were in my tummy we didn't know if you were a boy or a girl. My tummy grew bigger and bigger and one day we went to the hospital and after a very, very, very long time Mommy pushed and you came out and Daddy said, 'It's a boy!'"

"Did it hurt?"

"It hurt a lot."

"Were you bleeding?"

"A little."

I swear to God.

Kinship

(ITTY, USHA, FETA, AND THE TWO MOMMIES)

We're celebrating the life of my cherished grandmother's youngest sister, who died a few months ago at the age of ninety-two. With the exception of one sister, my grandmother and all five of her siblings lived into their nineties. I've always assumed I would live into my nineties too, and my sister, and my mother, and Faith by association. My father I've thought might kick off in his eighties—punishment for years of smoking, drinking too much on social occasions, and leaving my mother. It would be sad, but he would be suffering in the end—from a stroke or some other such thing—and so mostly we all would be relieved.

The memorial is an impromptu affair to be held at the Jersey Shore condo of one of my mother's first cousins. Most of my mother's side of the family will be there, sans rabbi, to share stories of Aunt Ada, show pictures, and generally have the good time she would have wanted us to have. Because of the public atmosphere of this side of the family, as well as some articles I recently have written, every last one of them knows Faith and I are trying to have a baby. Cousins come up and ask how it's going. One of my aunts grabs me and rubs me up and down, up and down as she shares stories of her proclaimed supersonic fertility.

"I drop eggs even when I'm menstruating."

But even more cataclysmic than the ritual of a fertile woman rubbing another who is trying to get pregnant is being surrounded by family, being in a home in which the photographs on the wall contain familiar faces, sometimes even my own. It seems welcoming for an embryo. This is your family. They embrace your will to live. They will hold you and cherish you so that you may implant. What is pregnancy anyway but an extension of family, a way for all of us to get our genes into the next generation, part of the vast circle of life and love.

A lesbian cousin in her seventies takes my face in her hands and announces, "I know you are pregnant."

I take that as a form of blessing, not necessarily a premonition but a wish, support, hope.

"I mean it. I'm not just saying it. *You* are pregnant."

My period is not due for another week. "We'll see," I say.

"You'll see."

Another cousin corners me whenever she gets a chance to reassure me hormone injections and egg harvesting are really painless, no big deal, easier than getting your period every month. *Nod nod. Wink wink.* Because we both know what kind of hell that is when you're trying to conceive. I may be gay, but I am part of the club! Someone else reveals that the aunt we are there to memorialize had great difficulty getting pregnant, and when finally she did it was only to lose the baby during childbirth.

"On her deathbed," someone whispered, "she cried out, 'Where's the baby? Where's the baby?'"

"You never forget," somebody else says.

Someone else shakes her head in agreement.

After bagel breakfast we drive to the beach and wheel my still-weary mother along the sands to the water's edge in a wheelchair with oversize tires, huge soft cushions, bumpers, and a seat belt. It is a wheelchair built for the sand. This Jersey beach, it turns out, is entirely accessible. As soon as my mother climbs out and takes a slow brief stroll under her own power, Faith and I and my mother's brother take turns trying out the chair. Someone dips a toe in the water, someone else collects shells, another lies back in the sand.

There are men among us, but on our way back to the boardwalk it's my job to push my mother along. In the beginning of my mother's illness I struggled to maintain a fifty-fifty divide of duties between myself and my sister, struggled not to feel like my mother's illness was my job, my responsibility, my life. I think now that trying to keep everything fair, everything predictable, everything the same was just a way for me to resist accepting the fact that she was sick. Most of struggle is just that, a resistance to what is. So now it's not that I like the situation, but it is what it is and at least I am here and able to care for my mother, I tell myself. It is a gift that I am close by. There is no more myth of a division of labor, no more effort to make the situation invisible. I am hers and she is mine. There is peace in that acceptance and a chance to let go slowly, in awful increments it would be more devastating to have to do without. I have always been the one in need of emotional rehearsals, these practice separations. That's how I do things: step-by-step, IUI by natural IUI. So all is as it should be. I am where I belong, pushing my ailing but joy-filled mother along the Jersey Shore.

At night we laugh until tears come at the names of our deceased relatives: Itty Gordon, Usha Ross, Uncle Yonkel, Feta Pacey. The names strike us as ludicrous, impossible. Just as the thought of a pregnant lesbian niece might have caused each of them to drop to their knees in laughter, or in prayer.

Not to give myself a *kenahora* or a jinx, because it is not like my saying so would cause or negate a future already in the process of becoming or not becoming, and as my aunt by marriage—second marriage, to be exact—said over the weekend, "There's no way not to be disappointed. You can't pretend you're not hoping, and why should you?" But anyway, not to jump the gun or anything, but last month I got my period on Day 25—early, I admit, but I did. And today is Day 25 and I still don't have it. That's it. That's all I wanted to say.

Day 28 is when I officially find out. That's the day I can have a blood test to determine whether or not I'm pregnant. It also happens to be my 40th birthday. The timing is akin to that of a film student's first screenplay or a very schlocky romance novel.

Day 26: Lots of depressing menstrual cramps that cause me to want to break up with Faith and move to a forest, where I'll build myself a cabin, grow my own chard, and raise basset hounds for the rest of my life. I would be the Unabomber without the bombs, a refugee from mainstream America.

That evening I dream that I've sneaked into my therapist's waiting room in the middle of the night and made a bed for myself on the sofa. She finds me in the morning and says, "I am way too nice to you."

Day 27: The cramps persist, and in response I conduct a vast Internet search of Clomid and ovarian cancer. Clomid + cancer. Clomid + ovarian + cancer. Clomid + risks. My Web search reveals all sorts of terrifying information regarding Clomid and ovarian cancer, including a *Boston Globe* article that states the FDA is considering asking its manufacturers to put a warning on the label: "May be associated with an increased risk of ovarian cancer." It's not the Clomid itself that may be associated with cancer, but what the Clomid does to a woman's ovaries increases her likelihood of developing the disease. Clomid causes ovaries to go into overproduction, speed-production, causes a woman to ovulate a year's worth of eggs in a couple of months. And that, apparently, is just the immediate effect. There's little information about its possible long-range effects. There are bulletin boards on several ovarian cancer Web sites (granted, a biased sample) filled with messages from women who have ovarian cancer and who prior to their diagnosis underwent infertility treatment. There's even a reference to a study that likens the potential long-range effects of Clomid to those of diethylstilbestrol (DES).

Disbelievers in the Clomid-cancer link suggest that it could be infertility itself that is a precursor to ovarian cancer. Maybe poor little Clomid just happened to be at the wrong place at the wrong time, hanging around with all these chicks and their loaded ovaries, and poor little Clomid was left with the smoking gun.

I take an Internet quiz to determine my risk of developing ovarian cancer without Clomid. Thanks to the dreadful combination of prolonged childlessness and my mother's diagnosis, it seems my risk for ovarian cancer is two times that of the general population. And Clomid, some assert, can increase that chance even more. On the other hand, the dastardly effects of Clomid don't tend to come into play until after three cycles, or three months of use. Therefore, I reason, two cycles—as in two Clomid-assisted IUI's—should do no harm.

The cramps come and go, ranging in severity from mild to moderate, like they can't make up their mind. My sister calls from Los Angeles. She has just gotten news that her amnio results are all fine, and she is feeling fearless, free of worry. For the first time she allows herself to wallow in the upcoming joys of motherhood.

"Just take Clomid the next time around," she says from the other side of the looking glass. "It's worth it."

And, though this cycle is not yet a definitive failure, I decide to try. I will try Clomid for two cycles and no more. I will have done my best, under the circumstances, and we can move on to our spare womb.

Later that evening, I tell Faith of my decision. She is concerned, doesn't want me dabbling in dangerous science, but knows a girl has to do what a girl has to do, if only twice.

"By the way, how are your cramps?" she asks.

That's when it first occurs to me. "You know, I think they're gone."

After dinner, Faith has an idea. She will lead me in an impromptu birthday fertility dance. She accompanies our made-up waltz with a made-up song she sings into my ear about babies and turning forty, about our life together, the kitty, and our life yet to come. The song ends with a kiss, and because it is already Day 27 and we are standing there together relaxed and free for the rest of the evening, because why not, I suggest we do a pregnancy test.

Faith usually forbids the performing of pregnancy tests until the day my period is due. She claims I am being mean to myself by taking them a day or week before. Of course, I have been doing them anyway, secret heart-wrenching tests long before my period, or days afterward, just in case. I figure she will say, "No, let's wait," and then I will sneak into the bathroom and do one anyway. But she surprises me.

"Okay," she says. "Let's do one."

Because I've performed about sixty-seven pregnancy tests in the last ten months, I know for a fact that negative is negative. Negative is unequivocal. Negative is easy. There is only one line. But positive can be anything. Positive can be the faintest hint or shadow or ghost of another line beneath the obvious pink reference line. So I blink my eyes to moisten my contacts and prepare to squint.

Faith examines the state of her complexion in the mirror over the sink as I aim a stream of urine into the small target on the test kit.

"I can't stand these dots," she says.

"They're not *dots*. They're freckles."

"That doesn't make them any better."

"Sure it does. Having dots is weird, but having freckles

is normal. No one calls freckles 'dots.' " I remove the kit from between my legs. A perfect puddle of urine fills the target.

"You have good aim."

"Thanks." Why tell her about all my practice runs?

I rest the test on the sink and try not to stare at it. Faith diverts her attention from her dots to the kit and yelps when a dark pink line starts to form.

"That's the reference line," I tell her, my heart rising and then sinking with a mighty crash.

We wait a moment longer. Nothing.

"You'll try again tomorrow," Faith says and returns to her facial.

But I don't believe it. I stare harder. Didn't I have cramps that came and went? Didn't I have those side to side pains my sister told me both she and a friend experienced the week before they missed their period? Didn't my excessively fertile aunt rub me up and down? And how about the cousin who pronounced me pregnant, blessed with the task of carrying the genes of our family into the next generation? I stare and stare until unbelievably I actually think I see another pink line form beneath the reference line. It's so faint, so light, so barely present that I'm certain I'm forcing myself into seeing things.

"Do you see the slightest line underneath the top one?"

My realist girlfriend will tell the truth. She looks down at the kit, studies it for a second, and announces, "Yeah, I do."

"You do?"

"Yes."

"You see a second line?"

"Yes. But it's really faint."

"Faith."

"What?"

There are no fireworks, no balloons. There is no orchestra climbing its way to a frenzied crescendo, only Faith, naive to the delicate workings of pregnancy tests, standing there mid freckle trying to assess my reaction.

"I'm pregnant."

"Do another one."

I repeat the procedure, all the while explaining the ins and outs of pregnancy tests, how one line means it's negative but any degree of another line, any half-baked, half-assed, whisper of a second pink line is positive. Positive! Positive! Positive!

"But it's so faint," she keeps saying. "It's just so faint."

We're supposed to be ecstatic, filled with tears of joy, and prayers of gratitude. Instead we're having a debate.

When the second test offers us the same scenario—one heavy dark line atop a light trepidacious line—Faith starts to believe it. We do a third. A fourth. Each time a little hint of a line slowly emerges under the reference line like a smile or an angel or the slight, precious finger-sweep of a newborn baby.

"Holy shit," my life partner and future coparent says.

I misread Faith's expression as shock, confusion, as if she wasn't aware that this has been our goal for the last eight months and instead sees it as some kind of biological faux pas, an accident of biblical proportions.

"Are you freaking out?"

"I just don't want to believe it until you have a blood test. I want to be sure."

"Okay."

"Before I believe it."

"Okay."

"Before we tell anyone."

"Okay."

We do a fifth test. There it is again, the little faint pink line smiling back at us. I can almost here it say, "Hello, Mommy. Mommies. Hello, *Mommies*." Blood tests be damned. We hop in the car and race off to tell my mother.

Happy Birthday!

(Canoes, Cell Phones, and River Magic)

Friday, Day 28, happens to be my fortieth birthday. It's like a made-for-TV movie, just entirely tacky and melodramatic that we actually might find out—for certain—that I am pregnant on my fortieth birthday. I mean, of all things.

In honor of this traumatic, very much divisible-by-ten occasion, Faith and I each have taken the day off from work to canoe three and a half miles down the Charles River to a little suburban waterfall, where we'll beach our rented boat on an embankment and have a picnic. The new fertility clinic is on the way to our starting point on the river. So a pregnancy test fits in perfectly with our adventure, and fifteen minutes after the 9 A.M. blood draw we are at a deli stuffing our cooler with picnic items. A half-hour after that we're comparing paddle strokes.

The day is so gorgeous and spectacular that if in fact this were a movie scene, the audience most definitely would let loose one big simultaneous groan, "Oh, right, off they go after seven months of potential infertility, this

FORTY-year-old lesbian and her girlfriend and their five percent chance of getting pregnant, paddling along a beautiful river on a sparkling, sun-filled afternoon in early June. Of course, herons and egrets line the riverbank; huge, cumbersome turtles float impossibly next to logs; frogs croak from lily pads; and wildflowers bloom among the cattails and willow trees. Yeah, right."

I agree. It's total schlock, entirely unbelievable.

Faith has brought her cell phone so that the nurse from the fertility clinic can call us with the results. The phone is jammed into a side pocket of the cooler, antenna withdrawn and waiting. Until the blood test Faith and I assumed I was pregnant. Five urine pregnancy tests couldn't be wrong. But after our visit to the fertility clinic and the nurse's noncommittal "We'll see" as she performed the official test, we are newly uncertain, cast again in premenstrual limbo. We row nervous little strokes, Faith up front, me in back, both of us gazing out from beneath the brims of our sun-shielding safari hats, blind and bewildered despite the gloriousness of the day.

"Don't overexert yourself," Faith calls from the bow.

Off to the right, three children and a golden retriever romp merrily in a field.

"Jesus," I say, because what else can you say on a day like this. "What a fucking day."

Because this scene in the drama that is our lives apparently is not schlocky enough, the phone rings just as we leave behind children and fields and civilization and round one of the most beautiful bends in the river. I lunge for the phone in time to receive a brief scowl from an older couple floating by us in the opposite direction.

"Harlyn?" the nurse asks in response to my urgent and unnecessarily loud "HELLO?!?

"SPEAKING!" I never say "speaking." But today, because I am forty, quite possibly pregnant, and in a boat I say, "SPEAKING!" really loudly like an old person.

"It's Joyce from the Reproductive Science Center…"

If this were a movie the other couple and all of the flowers and all of the wildlife would burst into song, fireworks would light the sky even though it is daytime, my deceased grandparents would wave from the clouds, and my mother's ovaries would appear before us all healthy and apologetic. But it's not a movie; it's reality, some unbelievably blessed day on a river in suburban Boston, some strange unexpected joyous moment after a long winter of gloom. And it's my fortieth birthday, for crying out loud! I hug my paddle to the boat so as not to capsize us, and listen as Joyce, the nurse, says "Congratulations, Harlyn! You're pregnant!"

Faith reads the expression on my face, and before I even have a chance to put the news into my own words shouts to the couple in the other boat, "She's pregnant! She's pregnant!"

They look relieved. We're not two corporate types who would rather be caught dead than in a canoe on a weekday without a cell phone. We're just two lesbians waiting to find out if we're going to be mommies!

"She's pregnant!" Faith keeps yelling at them until finally they smile and clap their hands and float away.

The nurse is instructing me to write down the dates of my next two appointments—more bloodwork, an ultrasound.

"I can't," I tell her. "I'm in a canoe!"

"Oh," she says. "Oh, my."

"And it's my fortieth birthday."

"It's her fortieth birthday!" she calls to someone. "And she's in a canoe."

I hear someone gasp in the background.

The instructions will have to wait. Besides, the flowers and the river and the turtles and the herons and the egrets all *have* burst into song. I *do* see my grandmother, and my mother, healthy and exuberant. There are rainbows and fireworks and kisses from each breath of the wind.

After the phone is back in the cooler, the couple out of sight, the river ours alone for the asking I whisper to Faith, "I'm pregnant."

She is staring at me, all teary-eyed and exuberant. The boat holds us still for a moment and then, in a breeze, off we go, two new parents in a tippy canoe.

PART TWO

The Trip

Baseball, Apple Pie, and Harriet and Harriet

People with children are so happy for you when you get pregnant. They are happier than you are, happier than they were when they found out that they themselves were pregnant. They are ecstatic. And while Faith and I have not yet fully digested the news, suddenly our place in the order of other people's lives changes. Women I hardly know confess to me the size of their breasts, vaginas, and perineum. Previously silent and disinterested men take Faith aside and engage her in long conversations about labor support, sleeplessness, and postpartum depression. We are offered onesies and changing tables and extra diapers.

"Oh, great! I've got a whole carton of 'twos' I can give you," a friend says.

"What are 'twos'?"

"They're diapers, silly."

Oddly, we no longer are gay, no longer alternative, no longer a minority. We are integral cogs in the reproductive machine, soldiers of evolution. We could be two zebras for all anyone cares. Because we are having a baby! We are going to become parents. We're going to become "normal" and identifiable. We're not quite Ozzie and Harriet; still, Harriet and Harriet is suddenly okay as long as a baby's in the forecast.

That I'm able to show up at my mother's apartment at 7:30 one evening and bless her with the news that I'm pregnant and witness her unequivocal bliss is not lost on me. The way things have been going it was equally likely we would not have been able to share this joy with my mother. Telling her I'm pregnant is not just *like* rushing over and giving her the biggest, best gift in the world. It *is* giving her the best gift in the world. It takes her mood and steals it immediately away from illness and liver tumors and chemotherapy. It kicks cancer in the ass, releases her from its grip, casts for us all a major blow against the enemy. We are living fully anyway. We are happy in spite of you, cancer. We have, for a moment, even forgotten you are here, and with that erasure have chipped away at the awful hold you have had on us for so many years. We have won a battle, if not the war. For the evening we are ecstatic and distracted and filled with joy, and you do not even exist.

My mother and Faith and I start thinking of names, guessing the gender of our small bean. We plan summers in Provincetown together, cross-country reunions with Carrie and her family, shopping trips for maternity clothes. On my mother's sofa this evening there is nothing but victory and joy and hope for the future. If I can get pregnant, then Mom can get well. The tide has turned. Suddenly anything and everything is possible for this family.

Each day for the first three months of my unbelievable pregnancy I still have to remind myself that I'm pregnant.

"You're pregnant!" I say to myself over and over again and each time it blows me away with much the same force. It's the same for Faith.

"Sometimes I forget," she admits one night.

"I know," I say. "Me too."

Simone has decided she wants to stay with us for a week. She is only four, but she's the most social and confident person I know. We compromise and decide on a visit that will last one night. Tory and Jay drop her off at the house in time for dinner. They've packed for her a small bag of toys, clothes, sippy cups, children's toothpaste, a tiny little toothbrush. We begin the adventure at 4 P.M. with a game of Cinderella (at her insistence she and I are the evil stepsisters and Faith is Cinderella). We move from that to Alice in Wonderland (starring Simone as Alice and me as a cross between the Mad Hatter and the Cheshire Cat). It takes a game of "Baby" to get our small guest to eat her dinner (i.e., Simone sits on my lap and I cut a turkey dog into little pieces, dip each one in ketchup, and place it directly into her mouth). After dinner it's Barbie, several books, three attempts to turn out the light, and finally sleep at approximately forty-five minutes past bedtime (a major victory, as far as I'm concerned). The next morning Simone's up at 7 calling me to come and resume Cinderella. The evil stepsisters have to make do on their own as "Cinderella" has decided to sleep in ("This overnight was your idea."). There's breakfast, a walk, a trip to the playground, and a new game, "Dance," which involves Simone twirling down the sidewalk to my constant applause. We make it back to the house and it's not even 10 A.M. At 10:30 I buckle and call Tory and tell her it's time to retrieve Simone so I can either take a nap or throw up. I tell myself it's because I'm pregnant, because Simone is already four and I didn't have the preceding years to build up to this, that I'm exhausted to the point of tears.

My mother reassures me that it's not the same with your own child, that with your own children you set limits and teach them to entertain themselves. She swears my eighteen hours with Simone in no way resemble parenthood. But I'm guessing there's as much truth to that as there is to the assertion that lobsters feel no pain when you submerge them in a pot of boiling water, that infant boys don't suffer in the same way a man would when their foreskin is lopped off. Still, I believe her because I have to. Because there is no turning back.

The first few months of pregnancy are no big shakes. Basically, you feel queasy or you don't. You lose interest in food or not. You gain a few pounds or lose a few. You feel tired and bloated, and your breasts take on lives of their own. Strangest of all, after twelve weeks of this, there you are with a third of your pregnancy behind you. And still some people have no idea you are pregnant! It reminds me of the morning I realized I would not be attending medical school. There I was sitting at a pond looking normal but with this really big secret, with this really huge event going on. Being gay and being pregnant is even more of an invisible payload. It's like, guess what! And then guess what again!!

The only queasiness I feel is in my head. I'm entirely and utterly under the spell cast by certain words: granola, coffee, Hawaii. I can go from robust to green in seconds just by thinking of lemon-flavored seltzer. I can force myself into shallow breathing by imagining sunflower seeds. The only foods I can stomach thinking about and eventually possibly eating are bread and butter and hard-boiled eggs. They sustain me for close to three weeks and then suddenly one afternoon the idea of them makes me want to gag. Cottage cheese works for a day. I never throw up. I just think about it. All day long I alternate between fantasies of introducing a child to the ocean and hanging my head out of a window to barf.

By the fourth month I'm just fat. It's a terrible state of affairs not to look pregnant, but instead to look like someone who needs to exercise more, or maybe get a better bra. Two women at work have an enthusiastic conversation about the Atkins Diet directly, and seemingly intentionally, in front of my office door. I don't want to make a general announcement about the pregnancy until after the amnio, just in case. But it's hard. Part of me wants to shout, "My breasts are huge because I'm pregnant, not because I've suddenly chosen to go to pot."

Despite my increased girth, despite breasts that enter a room and introduce themselves before the rest of me arrives, my weight is not much different. At our second prenatal doctor's appointment, before I have the breathtaking experience of hearing our baby's heartbeat superimposed upon my own, I'm shocked to find out I have gained only three pounds. I'm certain each breast weighs at least five pounds and my stomach another fifty. But there you go. And there is this heartbeat inside me, and it is steady and sure, and my God, it seems that I am pregnant. After all this time, all the obsessing and all the doubt and forty years of intensely contemplating my own navel, I'm going to be a mother. And Faith is going to be a mother. We *are* mothers. And with that realization everything changes. It's not that we're "so-o-o different now." It's that we're exactly the same, only the world around us seems to have suddenly transformed. I consider my own relationship with my mother—the inevitable, all-importance of it, the life and death of it, the shared soulness of it, and try to imagine someone feeling that way of me by virtue of having come to life inside my body, by my side. In some ways we are so ill prepared for that aspect of this adventure; that we are to become the most important people in the world to another being is so much bigger and so much deeper than the issues of money or privacy or cotton versus disposable diapers.

The sperm donor has all but disappeared from our experience of pregnancy and parenthood. It's incredible how absent he has become—he served his purpose and now has about as much to do with our daily life as my second cousins once removed. He is there, certainly, in biology and from time to time in thought, but the daily reality of parenthood and pregnancy have nothing to do with him or his semen. They have to do with us. Faith has taken over many of the chores. We eat well, plan doctor's appointments, buy books to help us learn about baby bathtubs and how to get an infant to sleep through the night. We save money and put up shelves. We clean the basement, rest our hands on my budding stomach. What does this have to do with anyone but us? It must be similar for adoptive parents. There is biology and then there is life, at least in the early years of a child's experience. It reinforces for me how intruded upon I would have felt by the presence of a known donor and his family. It is Faith and my and our family's baby. Maybe it all would have been loving and supportive and great, but it would not have been *us* and our commitment to each other. It would not have been as much our own shared and private joy.

The Cancer in Your Doctor's Eyes

My mother has become very ill. The remaining cancer in her liver has grown large enough to cause her debilitating fatigue, back pain, and a frightening amount of weight loss. At the second chemoembolization there had been only a fifty-fifty chance the doctors would be able to reach the renegade tumor with a catheter in order to infuse it with chemotherapy. And while today's follow-up CT scan reveals they did indeed reach it, the doctor's expression is somber, not at all victorious or celebratory.

My mother and I have become exquisitely adept at reading doctors' expressions, and this one is not at all good, despite the news. I feel my stomach sink. I feel my mother's stomach sink. I want to turn back the clock to some time before this moment, to stop everything right here. No more information, please. As if it is the awareness of cancer and not cancer itself that kills people and destroys families.

"It seems there's another tumor," the doctor says, despite my having sent him a telepathic message to shut up and go away.

Thirty minutes later a third chemoembolization has been scheduled, and my mother and I are sent on our morose way.

There are so many emotional stages of cancer, for the person who is ill as well as for his or her loved ones. There is the initial diagnosis and the belief that it all has been one big horrible mistake but treatment will happen any-way—Why not play along?—and then there will be remis-sion and the diagnosis, like the cancer itself, will fade into some distant dreadful memory that emerges every few months or so when it is time for a follow-up doctor's appointment, until the almighty five years have passed and one can consider themselves "cured." That is the eas-iest stage. You experience terrible trauma but live to tell and grow and appreciate life in ways you never thought possible. Next is the recurrence stage. In some ways, this stage is harder than the stage of initial diagnosis. This is the "you ain't getting off so easy" stage, the there-is-no-escape, the overwhelming, life-shattering "uh-oh" stage. From here on in nothing is going to be clear-cut or simple. Now you think not just about putting the cancer into remission, but keeping it the fuck away—an inning in which you already have one out. The stages that follow are varying shades of horror and hope, depending on the situation. Sometimes you move from stage to stage with-out even realizing it. Other times the transition comes along like a quick, sharp slap to the face.

The stage we are in now with my mother crept up on me long before our disgusting visit to the doctor with the somber face. It caught me one afternoon as I tried to encourage her to force down a little watermelon, Jell-O, anything. That's when it occurred to me that my mother hasn't been well in a long time, that there has been no recovery of energy and appetite in months, that her cancer is now symptomatic, and that she has not been in remission for well over a year. And then all of the rounds of chemo, with all of its strange and terrible side effects, and all of the times I've done her laundry or carried her heavy groceries from the car to her kitchen, all of the times I smiled when she joked that she still had not been able to return the copy of *Heather Has Two Mommies* she bought for us (our third copy) because she never could park close enough to the bookstore to walk in without tiring—all of it hit me. And then, *Shit*, I thought. *Shit. Shit. Shit.*

I remember thinking the loss of my mother somehow would be mediated by new life, by a child, by becoming a mother myself. And certainly this baby inside me in some ways feels like a gift from my mother—a gift along the lines of learning to walk, to paint, to fall in love with the sea—something she very gently and very continually encouraged me to do. ("Imagine my life without you.") While I cannot guess how I would be feeling now about this most terrible of stages in the course of my mother's illness if I wasn't pregnant, still. This is just too hard, sometimes too damn difficult to bear, despite new life, or maybe even because of it.

A gay man I know is getting a puppy for him and his partner to raise, a German shepherd. Before the puppy arrives in their life, he and his boyfriend will first spend a week on a lake in New Hampshire sitting in Adirondack chairs and drinking cosmopolitans. On their way home they will pick up the puppy and then spend the next week helping it to acclimate to its new color-coordinated digs.

Gay men know how to live. They make money, sip cocktails, and limit their caretaking to creatures with ten-year life spans who never have to go to college or pre-school. Don't get me wrong. I'm thrilled by this life inside me, by the mere miracle of growing a baby. The little bump in my stomach one day will smile, hold out its hand, and call me Mommy. It's what I want. It's just that cocktails, a summer home, and a clean, short haircut sound pretty damn good too.

Amnio Monday

(The Gender Sweepstakes, Spirit World, and Celestia of the Fourth Dimension)

I'm certain beyond a shadow of a doubt that I'm having a boy. I've known I would have a boy even before I became pregnant, and becoming pregnant has only strengthened that intuition. It started with my worrying I would be disappointed if I ever had a child and it was a boy. To prepare myself for that possibility, to prevent some horrific, embarrassing, shameful reaction upon learning that the life inside me was male, I told myself that if I ever had a child, it would be a boy, that life is like that, so there. And from that moment on I forbid myself to wish for a little girl, a daughter; I erased that pink-and-lace scenario from the realm of possibility and began picturing diapering, raising, and growing a son. Eventually I became so accustomed to the smiling, energetic little boy whom I would raise to be a smiling and kind man that I couldn't imagine myself with a daughter if I wanted to. I had fallen in love with him, with that part of myself that would mother a son. And then I dreamed of the little blond boy who smiled and giggled in my unconscious and announced, "I'm going to be your son." So I have no doubt at all. This boy is my destiny.

Four months and eight pounds later. Eight pounds! Not to harp on the weight thing, but I have never gained eight pounds in four months. Well, maybe in high school between my sophomore and junior years. I remember my mother taking me aside and then telling me, "Honey, you look a little bloated." She must have planned for weeks the most gentle way of telling her usually lithe oldest daughter that she was "filling out a bit" ever since she sprained her ankle and quit the track team. Still, it was like hearing I wasn't really human. Then there was freshman year in college. But all of that aside, eight pounds! The kid is barely six inches long, and it's not like I'm "showing." I'm just suddenly fat.

"How are you feeling?" people ask me.

"I'm fat."

"No, you're not," they say. "You're pregnant."

"I'm fat."

My waist has been replaced by jelly rolls, and sometimes I have to bend my knees in order see my own pubic region. I don't look pregnant; I just look bigger.

Finally, I run into an old friend who gets it. "You're fat," she says.

"That's what I've been trying to tell everyone, but they keep saying I'm pregnant."

"Well, you are pregnant. But you're also fat."

"Thank you."

"Sure."

This kind acquaintance explains that when a woman is pregnant her metabolism slows down and she retains more calories and fluid. Add to that the fact that her uterus expands and shoves her stomach out. I have no idea if any of this is true, but I feel so understood.

And then there's what I call the "just wait phenomenon." You run into another pregnant woman or a woman who has had children and she asks you if you're experiencing any one of a number of incredibly grotesque or distressing symptoms of pregnancy, and if you say no, she responds, "Oh, just wait!"

"Are you really sick every day?"

"No."

"I wasn't sick in the beginning either, but I threw up every day during my second trimester. *Just wait!*"

Or, "Do you have varicose veins and hemorrhoids?"

"Uh, no."

"Well, it's still early. *Just wait!*"

I want to ask these women why, if we never had a physiological or emotional thing in common prior to our pregnancies, would we suddenly become blood sisters? Isn't each pregnancy different? Still, they are better than the horror-story people.

"Oh, my God," the horror-story people tell you over dinner. "I know a woman whose baby was strangled by its own umbilical cord at seven months." Or, in passing at a wedding, "Did you hear about so-and-so? Everything was going along *perfectly* and then she miscarried in her thirty-eighth week. They say she should be able to come off of the antidepressants in about a month or so."

Our upstairs lesbians have done the unthinkable. They've started the endless rehab of their apartment, involving the demolition of walls, the sanding of lead-painted window ledges, and the toxic stripping and staining of doors. To a pregnant woman this is the equivalent of dropping an atomic bomb, spraying with DDT, taking in a family of locusts. This means war.

At the root of the problem is the fact that our upstairs lesbians do not even wear seat belts. How can you discuss the hazards of lead and benzene and oil-based paints to people who allow their bodies to be propelled around town—Boston, no less!—at fifty miles per hour unprotected? If they consider themselves and their year-old daughter immune to the environmental hazards of the world, why should they care about us and our unborn child?

At a wedding recently a bishop conducting the ceremony proclaimed, "Love is never just for two people." I want to scream this to our poisonous upstairs lesbians, but I have no idea what this has to do with anything. Maybe I just want them to look beyond themselves as they go about building bathrooms and rearranging gas pipes. What about us? Love is never just for two people!

An intervention is made on our behalf by objective members of our little environmentally conscious community. It is suggested that the upstairs lesbians do not need fetus-damaging toxins in order to step a few paces up the aesthetic ladder. Unbelievably, they concur. All use of sanding sealer—and anything else whose label contains a skull and crossbones—will be banished from the house. They may still use oil-based weapons, such as polyurethane, but the crew of non–English-speaking Yugoslavians employed to make the upstairs apartment more beautiful will be instructed not to use those in any common area of the house, such as the basement.

It's not enough, but it will have to do. When our waste

pipe explodes all over the basement the following week, I remind myself two-family life has, if nothing else, very worthwhile financial advantages.

Our amnio is scheduled for ten days from now, and we've decided that we want to learn the gender of our child. Premonitions aside (because when you come right down to it, I have absolutely no idea whether I'm having a girl or a boy), it is so huge, so altering and eternal, the fact of having a son or a daughter, that I find myself more nervous about that than the actual amnio itself. What, after all, is a needle in your uterus for forty-five seconds in comparison to the content of the rest of your life? I'm sure finding out the gender of your child at the moment of birth must be incredible and awesome, but I can't wait that long. Besides, it's not like finding out at sixteen weeks won't be a surprise in and of itself.

Then there are the amnio results. Maybe it is easier to worry about whether or not I am truly psychic about having a son than it is to worry that something is wrong with our baby.

I've heard that a spirit enters a new life at approximately twenty weeks in utero. Alternatively, I've heard that as soon as you feel your baby move inside you a soul has entered its small body. Either way, I'm waiting. As a result, I find myself adjusting my behavior in all sorts of weird ways: I'm kinder to strangers, more tolerant of my coworkers, better insulated from the madness of the road. All so that a really nice, mellow spirit might enter the body of my baby, rather than some deranged devil spirit without manners or remorse, as if the really nice spirit might be hovering over my belly at just the moment I turn on the Rolling Stones' *Black and Blue* album or flip some driver the bird and suddenly decide to make a beeline for another pregnant belly.

One night I dream about my paternal grandmother. She died when I was fifteen, and I hesitate to admit that I never really knew her, that I cared about her in a detached, abstract kind of fifteen-year-old way. In comparison to my effervescent, doting, and entirely overindulgent maternal grandmother, my paternal grandmother was shy and cold. Once when I was ten she refused to give me a cookie. God forgive me, these are the kinds of things I remember. Anyway, I dream of her and the next morning I confess to thinking, *Uh-oh.* It's not that I wouldn't want the spirit of my long-suffering paternal grandmother to inhabit my child, it's just that, well…why not my other grandmother? The next night I dream of B.B. King and am so relieved.

Though we've never discussed it, I think my sister and I are each secretly wishing to host the spirit of our maternal grandmother. I don't know which is worse, that I would choose the spirit of one grandmother over the other or that my sister and I are competing on such a spiritual and supposedly holy, Zen-like level.

"Nana's spirit is mine!"'

"No, it's mine!"

I decide to call Carrie and tell her this crazy, embarrassing idea, hoping she'll laugh and say, "Don't be ridiculous, we'll each have wonderful children," or, at the very least, "Everything happens for a reason." Instead she confidently and unhesitatingly explains that a few years ago a psychic told her that one day she would give birth to the spirit of her grandmother.

"Which grandmother did the psychic mean?"

"Nana Betsy."

"How do you know?"

"I just know."

"It could be Nana Goldie."

"I don't think so."

The air across the wires sizzles and smolders until one of us finally changes the subject. I imagine our grandmothers somewhere in heaven rolling their eyes and grimacing. In the land of karma and spirit, this is so not the point; in fact, it's probably earning each of us like a thousand extra lifetimes on earth.

As the amnio approaches I begin to question our decision to find out the sex of our child. On the other hand, Faith becomes more resolute than ever. She says things like, "Only six more days!" and "Next week at this time we'll know!"

It just seems that if this is my once-in-a-lifetime pregnancy experience maybe I should go whole hog and do the "It's a girl!" or "It's a boy!" thing. Besides, once you know the sex of your baby, what then? What else do you think about but his or her future health and happiness? And that would be just way too much to take on right now. Isn't it better to obsess over genitalia rather than genetic diseases or psychiatric difficulties?

To make matters even more confusing, without even knowing yet that I'm pregnant, our secretary blurts out one day, "I never even care when a baby is born if I already know its sex." Granted, this woman voted for George W. and refuses to see foreign films as she "didn't pay eight bucks to sit in the dark and read." But still, it's true. The world outside your closest friends and family is focused on only one thing: guessing the gender of your child. Face it, pregnant women are a community's Super Bowl.

A pregnant friend of mine who chose to learn the sex of her child at eighteen weeks' gestation says strangers on the subway and acquaintances at work are shouting genders at her all the time. Once she was eyeballed by an old woman on a street in Manhattan.

"You are definitely having a boy," the woman announced.

"Actually, it's a girl," my friend told her.

"No, it's not," the woman replied.

Now my friend just smiles and keeps on going.

My sister has decided not to find out her baby's sex, and part of me wonders if this is just one more way in which her pregnancy will be more authentic than ours. Maybe knowing the gender of your unborn baby takes something away from the miracle of childbirth, just as being inseminated with the frozen semen of a stranger, by a stranger, while your significant other is at work takes something away from the romance of conception. I try to imagine that moment—a baby emerging from inside my body, Faith by my side, a doctor saying, "It's a…it's a…it's a…" It seems so real and huge and meaningful. I worry the ultrasound technician will not make so much of a deal about the delivery of that information at the amnio, that the news will be off-handedly slipped into a tally of fingers and toes. "Femur, ulna, radius, heart, penis, bladder, brain." Maybe it's just that whenever we find out I want fanfare and excitement, balloons, at the very least the lighting of a candle.

But because I'm forty and life already has been chock full of enough surprises, and because I would love for my mother to be able to revel in imagining her firstborn mothering either a son or a daughter, Faith and I agree to find out at the amnio as planned. To appease myself I decide to ask that the news be delivered deliberately, gently, that we at least are granted a moment to pause and catch our breath, to picture ourselves the mothers of a son or the mothers of a daughter before a needle is inserted into my womb and we are hustled on to the next thing.

It's August, and my spirit-stealing sister comes for a visit. It will be her last visit east until after her baby is born. She's here to help our mother through chemoembolization number three, which is scheduled for the week before my mother's annual foray to Provincetown and three weeks before my amnio. The plan is to get my mom through the first tenuous week of recovery in Boston and then to gently, carefully, and without causing too much pain or discomfort, transport her to Provincetown, where we all will convene to continue her healing process—a beautiful and wondrous act that somewhere inside I believe will cure her of cancer altogether. On our side will be the dunes and our unborn babies, and the way the sun casts shimmering rays of light like jewels or crystals across the bay.

Carrie arrives from Los Angeles all big and round. Cars stop for her when she crosses the street. Cashiers smile as they count her change. Basically, everyone promptly diverts their attention from my suddenly barely noticeable tummy and great big breasts to my sister's protruding abdomen and flattened navel.

"You're definitely having a boy!"

"It's a girl!"

Just as expected, it's like she's the Kentucky Derby and people cannot wait to place their bets.

That said, I admittedly have grown accustomed to being the center of familial attention, and with the arrival of my enormous sister I am hardly noticed. At seven months her pregnancy totally steals the show. It's Carrie this and Carrie that. To make matters worse, all day long my sister rubs her stomach and announces each time she has a Braxton-Hicks contraction. She will walk only a block or two before slowing to a halt and wondering aloud whether or not she is dehydrated. We are kept well-informed of her baby's every movement and lack thereof. Her husband abandons hard-won parking spaces in order to quickly retrieve the car so that he may pick her up and drive her the block or block and a half that suddenly stands between her and fetal health.

In response I find myself purposefully walking two miles a day, just to prove being pregnant doesn't mean you have to become a total wuss. I carry groceries and one day even rent a kayak, which I defiantly paddle across the Provincetown bay and back. Degrees of gestation aside, the differences in our experience of pregnancy are blaring and loud, symbolic of all the other ways in which we are dissimilar. Given a lifetime of contrast in the choices we have made, from education to relationship to career, this is perhaps one of the first things we have had in common since being raised in the same house by the same two adults, and our integration of it into our lives is as different as night and day.

It would be easy to blame Carrie and my differing pregnancy journeys on the gender of our spouses—that the whole seed-in-the-belly thing makes Kevin more intrinsically connected to Carrie's pregnancy than Faith is to mine, that it is the fact of his sperm having penetrated her egg membrane that explains why he is all strut and swagger, shuttling Carrie from door to door and never letting her walk more than a few yards unassisted—while Faith still sometimes forgets that I'm pregnant and the other morning over breakfast asked if I would like to learn to parasail this summer. It could be because Faith's genes have nothing to gain that she has not posted all of our ultrasound photos on the Internet. Or it could be that my sister and I simply have chosen different spouses, number of orifices aside. Kevin is compulsively in charge, and Faith is in parasailing, albeit adoring, space.

Somewhere during the course of my pregnancy it's gotten harder for me to face my mother's illness. I'm sure some of this is due to sheer exhaustion, that working full-time, preparing our household and each other for the arrival of our new family member, and climbing the mountain (doctors liken the physical exertion of pregnancy to climbing a mountain, some gorgeous peak out in the middle of nowhere; I'm not sure I agree, but it sure garners you empathy, makes it all the more difficult, and some days even impossible, to take on the heavy, horrifying task of taking care of my mother. I focus on the physical responsibilities— driving her to the doctor, cooking her meals, walking her dog, doing her laundry—and claim I can't do them anymore, that she must move into our house during the few weeks of her recovery, because maintaining two households is too much for me. It's true, but it does not take into consideration the other reason for my sudden inability to deal: terror. For some reason I have even less of an ability to manage the anxiety associated with finding out the result of a CT scan, with watching my mother limp about the house, too tired to climb the stairs to her bedroom more than once a day. I'm less insulated than ever, each emotion lies raw upon the surface. If my body is climbing a mountain, then my emotions must be floating across the Atlantic on a small inflatable raft.

I convince myself that if only my mother would consent to staying with us, everything would be better, just like during the blizzard that wasn't. She could sleep in the extra room, add her laundry and meals to our own. We could keep her company. And when she was resting, we would be able to go about our business. I imagine that despite her illness we would be afforded some semblance of ordinary life and that all of our mundane day-to-day conversations and activities might serve to distract her from the all-encompassing experience of her illness.

But she refuses.

"It's not like we live in a freezing cold barn and get drunk and fight every night," I tell her on our last evening in Provincetown. She is due to return home alone tomorrow and still cannot stand long enough to cook herself a meal.

"No," she tells me.

"I could cook for you and do your laundry."

"I'd really rather be in my own home."

For the last five years that has been enough to keep me shuttling from my house to hers, from my life to hers. My mother is sick. She needs to be comfortable. How can I complain about giving things up to take care of her when I am lucky she's here? But now it's different. I'm on my raft, clinging to the need for some joy at this time in my life, the need for Faith and I to share the last months of our childless union. Rather than sleep alone in my mother's pull-out bed, I need to be with Faith, so that I can lay her hand upon my ever-changing belly before we fall asleep and in the morning as we wake. I want to care for my mother, I do. But I also need to reserve some room for hope and happiness.

Faith worries that insisting my mother move into our house would deprive her of one of her few remaining comforts, her own home. I think it will help me to be a better caregiver. Rather than exhaust myself to the point of resentment, I will be able to give and take and ultimately grieve. Running around as I am is pissing me off, and that makes it all too easy to rest in rage and not experience the vast sadness of the situation. Besides, we live only four miles away. We have the space. We enjoy one another's company. There is a porch and a yard, and *really*.

My sister, worried herself at not being able to assist in future recovery periods, agrees.

"Come on," we plead with our mother.

Our mother refuses.

"We'll have a good time," I say.

"Yeah," my sister echoes. "You'll have a great time."

My mother smiles a sad smile. She's taking care of us, or so she thinks, by not burdening her pregnant daughters with her illness. Though it's far more of a burden to worry about her alone in her house, we have no choice but to let go of the battle. If being alone in her own home is what she needs to do to survive this, then so be it.

Upon returning from Provincetown, Carrie and I alert our mother's closest friends, "She's home alone and not doing well." A schedule is planned for each night of the week, meals are arranged, visits mapped out, all without telling our proud and independent mother.

On Saturday my mother calls me and says, "Mary

stopped by with dinner." See, you don't need to worry about me.

"That's great."

Faith and I are on for Sunday night, her friend Jeane for Tuesday, little does my mother know.

Amnio Monday. Our date with destiny is not until 1:15 in the afternoon, but I'm up at 6:30 A.M. with the day off, so there's really nothing to do but clean incessantly—as if rather than returning home with information we will be bringing home the baby him or herself, and things better be spotless! I scour the floors, the light switches, the piano. I find sawdust on top of a door frame from when we moved in three years ago and a Popsicle wrapper beneath a sofa cushion. These just spur me on, and soon I'm digging around in the quarter-inch gap between the stove and a cabinet, crawling around under the dining room table with a Dustbuster to suck up whatever the vacuum can't reach. When there's nothing left to dust or sweep, I hang a painting in the bathroom, wipe down the kitty litter box, and prepare to repot a plant.

At 8:30 A.M. Faith emerges from the bedroom puffy-eyed and groggy, 33⅓ RPM to my 78. She eyeballs the surgical gloves and mask I am wearing to enable me to clean without disturbing critical aspects of our baby's development, looks at the clock, the bucket of dirty soapy water, my outfit (underwear), and her watch, and heads for the bathroom.

My domestic frenzy takes me all the way to 10:30, and then I have absolutely no idea what to do with myself. In just a few hours we will know the gender of our child! In just a few hours a foot-long needle will be inserted through my skin and into my uterus! In just a few hours our baby's limbs will be visible, his or her tiny fingers and toes and ears and eyes, and hopefully all will be present and accounted for, hopefully we are not in for any surprises. I water the remaining plants, make a few phone calls, shave my legs.

Faith is relatively calm. With all of the anxiety I carry, there really is no reason for her to experience any of her own. In fact, I probably carry enough anxiety for the entire neighborhood. If you went from door to door you likely would find house after house of calm children and adults, moving around slowly and smiling peacefully. Thank goodness for that nervous woman down the road!

Faith peeks into the shower, where I'm furiously trying to remove every hair from my thighs down, takes my face into her hands, and shrieks like a *bubbe* from the old country, "What good is worry? What does worry do?"

"Nothing," I say into her clenched hands.

"That's right! Nothing!"

It doesn't help, but it makes me laugh all the same. I laugh nervously, careful not to lose my grip on the razor as well as on the anxiety—which I am convinced is keeping evil at bay—careful not to upset the balance of the entire neighborhood.

Time continues to crawl a slow and tortuous drag all the way to noon, and then it suddenly flies and we are racing to brush our teeth, find my health insurance card, keys to the car. Moments later I am lying on my back on an examining table, Faith by my side. Surrounding us are a female radiologist, our obstetrician—also a woman—and a female ultrasound technician. There is not a man in the room, except perhaps for the one inside me, and that causes me an instant of pride, of estrogen-laden empowerment. Here we are, five women with tens of thousands of dollars worth of technology, a pregnant womb, and the ability to see inside it and embark on the journey of a lifetime. We are the future. We are the present. We are science—hear us roar! Suddenly, asking that the news of our baby's gender be delivered gently and poignantly seems mushy and wimpy, like women have achieved so much in this world—MDs and Ph.Ds, technical wizardry, insemination without the presence of a man—how dare I set us back a hundred years and ask to be treated like a delicate flower. I'll take the news like scientist, like an astronaut, like a real guy.

A choice spot on my lower abdomen is swabbed with beta-iodine by our ob-gyn as the radiologist glides the ultrasound wand across my stomach. The technician stands by, arms folded. Once again, the movie playing on the ultrasound monitor is *Harlie's Uterus*. Only this time we are greeted with substantial grainy blobs of gray and white and black that somehow, impossibly, are a fetus—not an embryo, but a whole little child-to-be. When a bit more pressure is exerted on the wand, the shadows become part of a tiny skeleton and I can make out an intricate spinal column, a very detailed hand, a tiny skull.

Everything from that point on happens so quickly I hardly have time to take a breath, to ask a question, to revel in the wondrous miracle unfolding before our eyes. There is a person living inside of me! I am a house, a home, a mother! I don't even remember my fear of the last four months that I would cough or sneeze during the amnio and cause the needle to shift dangerously near to our baby. I turn to Faith. What is she thinking? What is she feeling? Where is she?

"Don't look at the needle!" our doctor commands with a stern smile.

The needle, though long as heck, is far less interesting than the monitor. I don't care about it at all. I was just trying to find Faith, whose whereabouts I suddenly couldn't remember, even though it turned out she was standing behind me, patting my hair and stroking my forehead. I return my attention to the movie, the show of shows, soundtrack by Patti Smith: "Every woman is a vessel, is evasive, is aquatic…"

Despite the rapid pace of swabbing and coating my stomach and locating the fetus, there actually is time to forget about our baby's gender. I imagine this is what happens at childbirth: You are so overcome by the process of birthing a baby that once he or she is all pushed out and lying on your stomach, you forget to care, or perhaps simply no longer do. I think that now is like that, that seeing a fully formed being inside me is so incredible and unbelievable and amazing that it hardly matters at all what gender the little being is, as long as it's healthy.

That is until the radiologist, gleeful at the ease with which anatomical information apparently is available, proclaims, "That baby is spread-eagle. Do you want to know?"

Wait!

I want to stumble, to hesitate, to stutter, to go back to the moment when it didn't matter, when I wasn't even thinking about "he" or "she," but "Yes, yes, yes," Faith says. And because I am awestruck, on my back, instructed not to turn my head an inch to the left, because I want to know and don't want to know but want to know even more, I don't say a thing, just wait the long fraction of a second it takes for the words to form on the radiologist's lips.

"Well," she says, "it looks like a girl."

And then for just a split second, just a small silverous sliver of a moment, the earth stands still. When it starts turning again, when the fact that, yes indeed, I am going to be a mother, that I am going to have a *daughter*—that I already have a daughter—releases its hold on my heart and disbelief turns to belief and back again a few dozen times, only then do I marvel at how wrong I was, how hugely and immensely wrong. It's a girl. A girl. We have a little girl.

"I knew it," Faith says and proceeds to inform the entire room of the details of my false premonition. The women of mission control don't seem that interested in my dream of a little blond boy or the overwhelming feeling of maleness I insisted was coming from my womb.

"Don't decorate the nursery yet," our obstetrician warns.

"No," the radiologist interrupts her. "This one is pretty clear."

Our little girl, spread-eagle. She wants us to know: "Look, Mommies, no balls!"

Gender is put aside for the moment as the needle is inserted and three vials of our baby's—our *daughter's*—urine is removed for all to see. It is a sweet, precious, fragile shade of yellow, such unbelievable evidence that she is already in there moving and peeing and having a good time. Seconds later the needle is extracted, the radiologist holds open the door for our obstetrician, and the two of them are off and gone to the next thing, leaving us alone with the technician.

As procedures go the amnio is nothing, no worse than a blood draw, perhaps even less uncomfortable. Besides, there is so much else to look at and think about, and holy shit, it's not like waiting for CT scan results. It's positive. It's glorious. It's life.

Unfortunately, the ultrasound technician is in an entirely other mood than Faith and me. She is not part of a groundbreaking five-woman *National Geographic* expedition. She is at work on a Monday afternoon. This is so not one of the most momentous experiences of her life. Blandly she pushes the wand back and forth, back and forth across my stomach. *Blah blah blah blah blah.*

Rumor has it that if this detailed ultrasound is uneventful, the chances are better for good amnio results, that some of the major horrible things detected via the amnio can be forecast by certain distinguishable anatomical anomalies.

"How does she look?" I ask, the "she" catching on my tongue.

"She looks good," the technician answers in a tone one might use to comment on the weather. Looks good. Sun's up. Moon's down. Could change. Who knows.

I wish she were more enthusiastic. Still, Faith and I tell ourselves later that if anything had looked questionable she likely would have called the radiologist back inside to take a look.

The technician is kind enough to print for us a series of baby photos before she instructs me not to lift anything heavy for the next few days and then sleepwalks away.

Alone in the examining room, Faith and I are able to look at each other for the first time since I was commanded to keep my eyes on the screen. I'm not sure how we're supposed to react. If this were a movie, the husband and wife would start weeping. Maybe he would drop to his knees and give thanks to the miracle inside his lady's womb. Maybe she would start sobbing and bury her head into his starched white dress shirt so that he—a manly tear in his eye—could hold her there for a prolonged tender moment. Music would come up. A trick of cinematography would superimpose a faint image of the baby upon its two weeping parents. Of course the baby would be smiling broadly, gleefully clutching her umbilical cord like Dick Van Dyke did his umbrella in *Mary Poppins*. But here, back on earth, in this sterile exam room, our life together has been delivered into something wholly new and unimaginable, and not just because she will never ever again believe any of my intuitions. What do we do? What do we say? At last the words come.

"Wow," I say.

"I told you so," Faith responds.

Amnio waiting period. Anne Heche is spilling her newly married guts to Barbara Walters, going on about how screwed up her childhood was, how in response to horrors suffered at the hands of her clergyman father she resorted to becoming two different people: herself and a woman from the "fourth dimension" named Celestia.

Well, who fucking cares? The entire lesbian nation knew that relationship was doomed, that it was a crock of Hollywood shit. And what does this little pseudo heart-to-heart do but add fuel to the fire that lesbianism, even a brief three-year visit to the Isle of Sisterly Love, is the result of a weird-ass childhood. Still, my girlfriend is glued to the television.

"Please watch this with me."

"No!" I hate this celebrity manipulation. I hate television. I hate Faith for watching.

The next morning Faith, who is at home on school vacation, calls me at work to tell me details of the show.

"She had sex with her father every day on all fours. She says all she has is her memory to support her because so many people don't believe her."

"Oh, for crying out loud," I say from the psychiatric hospital where I make a living thanks to the fact that this kind of thing really does happen, thereby screwing people up beyond words, far worse than making them Oscar nominees. "I don't give a rat's ass!"

"Are you in a bad mood?"

"No!"

"Are you worrying about the amnio?"

"I just don't give a shit about Anne Heche."

Besides, at four and a half months I've already gained ten pounds, and pseudo-lesbian mental illness or not, that woman is just way too skinny to even think about.

"All I want is for my stomach to poke out farther than my breasts so that I look pregnant and not like a Barbie doll. I'd say that's a little more important than Anne Heche getting fucked on all fours by her father."

"I'd say," Faith says.

I can feel her rolling her eyes.

I dream I see the baby moving inside me: a hand pokes out here, a foot there, little fingernails push through my flesh like pinpoints. "She's going to tear her way out," I say in my dream, to no one in particular. It's as if the baby is covered in a thin sheet, like a child over whom a blanket has been thrown, and each of her little features are clearly identifiable beneath the sheet that is my skin—an enshrouded little hand, a foot. But she protrudes so dangerously that she is likely to burst through at any moment.

At this point in a pregnancy you don't want the baby to

come out, and so, according to my psychoanalyst mother, many women dream anxiety-ridden dreams of babies somehow slipping out of them too soon. I think she might be right. The other morning in half-sleep as I got ready for work I even thought I saw a little leg coming out of me. Whoa, I thought, stay in there!

My sister, being closer to delivery, reports having dreamed that she gave birth to a gerbil, which she had to store in a bowl of hot fudge until it was big enough to put in a Baby Bjorn.

Initially, waiting for the amnio results is easy, thanks to my mother's cancer. The results will take two weeks to come in. During the first week my mother is scheduled for a post-chemo CT scan, the results of which will tell us whether or not chemoembolization number three succeeded in destroying the tumor, whether or not the remaining tumor has grown dangerously larger, and whether or not the cancer has spread beyond her liver. Therefore, I can bounce from anxiety to anxiety as the spirit moves me. Worried about the amnio? Why not freak out about your mother's cancer? Your mother's cancer got you down? Why not run through a list of all of the things the amnio might confirm is wrong with your child? I alternate so rapidly from one stressor to the other that one day it occurs to me I'm actually relaxed.

"I think they've canceled each other out," I tell Faith with a relieved smile.

"Yeah, right."

As the CT scan appointment approaches I forget entirely about the amnio and ruminate instead over the horrifying state of my mother's health: daily nausea, vomiting, stomach pain, an inability to eat more than a couple crackers at a time. None of this happened after the other two embolizations. Without admitting it aloud to one another, we all fear the same thing, that the cancer has spread, that these are symptoms of cancer rather than side effects of chemotherapy, and that my mother is far too weak and debilitated to undergo further treatment at this point.

As I am homebound post-amnio, a close friend of my mother's accompanies her to the hospital for her 2:30 appointment. With any luck I'll hear from them by 5. At 4:45 the unimaginable happens: My mother calls with good news. The cancer has not spread. The tumor is dying. The remaining tumor has not grown. And all of that gastrointestinal distress was simply gastritis, caused by one of the painkillers she was taking. With a prescription for an antacid in her pocket, she is sent off and told she does not need to see a doctor for two months!

Our exhibitionist daughter lets loose a butterfly tickle to my side: "Right on, Grandma!" she shouts from deep inside my womb.

Purple Thrift

Whosoever has lived in Chicago as a bohemian or a student or an artist, as an urban vagabond or a lesbian, an immigrant or a costume designer, a penny-pinching retiree or a flannel-clad grunge; whosoever has visited a friend or a relative in Chicago—the city itself, not any of the outlying suburbs or subdivisions—and found himself or herself in sudden need of an affordable jacket or sweater; whosoever has no qualms about purchasing used T-shirts or jeans or baby clothes, who knows that most children's clothing is worn only once or twice and that once washed in Dreft or Ivory Snow is restored to the original full-price condition it was in at Gap Kids or Old Navy or the Children's Place; whosoever is willing to forgo prohibitive Jewish folklore that implores new parents not to buy clothing for a new baby more than a month before its due date, lest fate be offended and retract said baby's good health; whosoever purchases used baby clothes at an enormous discount an ominous five months prior to her baby's due date anyway—before even the amnio results are in—she shall never hesitate to enter the Village Thrift.

In Judaism there are a lot of things you are not supposed to do. You're not supposed to attend a graveside funeral service if you're pregnant. You're not supposed to eat pork or shellfish, or mix meat and dairy on the same plate. You're not supposed to wear leather on Yom Kippur, name a new-born after the living, or leave a foreskin on a penis. You should never mention the dead or the dying or the unfortunate circumstances of another without also muttering incomprehensible Yiddish phrases to prevent said misfortune from happening to you. Similarly, you should never mention your own *good* fortune without muttering something incomprehensible in Yiddish so that you don't jinx yourself and cause all your good fortune (past and future) to sink like the *Titanic* into an ocean of doom and gloom. And if you are pregnant, you should never, *ever* buy things for your baby until he or she is born or, at the very least, you should wait until a month before his or her due date.

Enter the Village Thrift, the enormous, breath-stealing used clothing store a few blocks from Faith's mother's house in Chicago (to which we have traveled for the weekend), aisle upon aisle of used (some a lot, some hardly ever) baby clothes, the most expensive item of which is three dollars, the least expensive ninety cents. To top it off, for some reason I cannot begin to imagine, the weekend we are in town everything is half-price.

So what is a "financially conscious" Jewish lesbian who has just found out the gender of her unborn child to do? Shake her head devoutly and visit the Museum of Science and Industry, or borrow a duffel from her mother-in-law, say a *kenahora,* and spend the rest of the afternoon going through every article of infant and toddler clothing the Village Thrift has to offer? I ask you.

Without a doubt, I have the blessing of every one of my deceased *bubbes* and *zadies,* even the kosher ones. We are a crew of motley, make-it-up-as-you-go Jews. My uncle Freddie, who at eighty-eight was known to wolf down an entire club sandwich from the deli counter at Shop Rite before reaching the cashier to pay only for a quart of milk and a loaf of bread, is applauding somewhere in heaven. My maternal great-grandmother, who drank beer and cracked crabs in the basement certainly would not blink an eye. My great-aunt Lena, who ran a brothel in New Jersey, and her son, a notorious kleptomaniac she often saved from incarceration by reminding the arresting officer of his visits to her house of ill repute—well, neither of them would give this a second thought.

No, it's only my contemporary musical-comedy girl-friend who expresses concern.

"They say you're not supposed to buy anything this early," she says, wide-eyed with panic.

"You didn't even have a bat mitzvah."

"It's a jinx."

I gently remind my suddenly orthodox partner that we already have taken fate and tradition into our hands and twisted them into an unrecognizable pretzel by virtue of our matching gender and my having been impregnated with anonymous, store-bought, mixed-race, goyish sperm. Therefore, it's pretty unlikely the gods are going to zero in on the baby clothes thing when making a recommendation for our family's future health and happiness.

"Besides," I tell her, "in case you forgot, you're a musi-cian and I'm supplementing my incredibly enormous aca-demic salary writing short stories—we need all the bargains we can get."

Faith acquiesces. Though it's less my argument than the racks of vintage T-shirts that distracts her from her anxiety. We meet at the register three hours and twenty-five dollars later. In our possession: ten T-shirts and thirty different articles of clothing for a little girl who as yet does not have a name.

With three more days to go in our amnio waiting period, I take stock of why we have opted to have an amnio in the first place. Is the purpose of the amnio to offer us information so that we may decide whether or not to terminate this pregnancy? Is it to offer us information so that we may be medically and emotionally prepared for possible disaster? The answer is less clear now that each morning when I wake there is a slight but significant lump on my left side, a very small girl curled up for the night, now that we have seen our baby slip one hand behind her head and the other into her mouth, now that there are rare but distinct little butterfly flutters from inside of me, a little girl twisting and turning and sucking her thumb. While I will forever support a woman's right to choose, I cannot imagine terminating this pregnancy at this point in time—at four months—regardless of the amnio results. I had thought I would be able to, that *we* might be able to. Faith shakes her head, of course not. It seems we already have become parents. And that makes the waiting even harder.

Tory says it's time to enroll our unborn daughter in preschool. She says it may seem too early, her being a fetus and all, but that actually it's not early at all. In fact, in all likelihood, we will be put on some sort of unborn wait list. It seems crazy, but Tory is the mother of a four-year-old, so she must know.

I find out the phone numbers of three highly regarded preschools. One is housed on an organic farm with chickens and goats. The gay-friendly staff grow pumpkins in the fall and tulips in the springtime. The children learn all sorts of preschool things like painting and drawing and skipping and how to nap at a designated time and poop in a toilet. They get to pet the goats and feed the chickens. They learn about flowers and spend tons of time outside. There are tears in my eyes by the time I've finished dialing the number, thinking of our hopefully healthy little girl playing one day among animals and hay.

"I'd like an application," I tell the warm woman who answers the phone.

"I'm so sorry," the woman says, truly remorseful about the news she is about to deliver. "But we're done accepting applications for next fall."

"No, it's for the fall of 2005."

"What is your child's date of birth?"

"She hasn't been born yet. I'm still pregnant."

The woman pauses, snickers, then laughs out loud.

"Day care!" Tory tells me later. "It's time to enroll in day care, not preschool."

Hesitantly, I call the day care center at the hospital where I work. Sure enough, our unborn daughter is wait-listed for enrollment seven months after her due date.

It's Monday morning, and I cannot wait another minute. Two weeks have passed since the amnio, and I've been pretty darned patient. Sufficiently distracted by disease and uncharacteristically optimistic about the results,

I haven't called the office once. I have not found out our doctor's home telephone number or harassed her into getting us the results early. I have not consulted a psychic, a Ouija board, or my therapist. I've been good. Very good. And now it's time for me to know.

I try to compare the anxiety I'm feeling as I dial the doctor's number with that I've felt when calling to find out the results of my mother's various cancer tests and biopsies, with that associated with waiting for medical test results of my own, with opening up the envelope that contained my SAT scores, my MCAT scores, a rent increase. But there is nothing in my emotional history that comes close to this, the apprehension surrounding life-altering information about someone so close to me—living inside me—whom I can't imagine living without, but whom as of yet I don't even know. It's someone else's future and my future all rolled up into one grand experience that defies comparison. I'm reminded of getting off a plane in India and venturing into Bombay at midnight with only my western mind for a reference point. To fully comprehend some experiences you must let go of everything you know, everything that has gone before, and let the landscape of your new world define itself. Perhaps we should do this always, with each new day and moment of our lives. But we don't. We compare over and over again, this to that and that to this. So moments like this, waiting for amnio results, are good lessons. They jar you from the habit of comparison, which tempers new emotions and experiences, cloaks them in familiar wrapping.

"Press three to leave a message for the nurse."

"Hi, this is Harlyn Aizley, and I had an amnio on August 27. I know it's exactly two weeks today, but I can't stand waiting any longer and was hoping you could let me know when the results might be in. Thanks so much." And then I go back to work.

It's easy to forget I made the call, some kind of magnificent denial that enables me to perform psychiatric intakes and maintain research records, all with blissful, ignorant, the-ball's-in-their-court calm. Until the phone rings. And then I slam closed the door to my office and futilely attempt to catch my breath enough to utter a "hello" into the receiver.

Three phone calls later it's the nurse.

"Harlyn, I can't believe no one called you."

In the seconds it takes the nurse to inhale a new breath my mind races. *Why?* Why should they have called me? To tell me good news? Bad news? Are things so bad that we should have known earlier? Are things so good that we should have been offered relief sooner? Was someone too afraid to tell us? Good news. Bad news. Good news. Bad news.

"Now, remind me, did you want to know the baby's sex?"

Jesus, I thought that was settled. "It's a girl?" I squeak.

"Yes, you're having a 100% chromosomally healthy baby girl. Congratulations."

I'm smiling because I think I'm happy, but what does "chromosomally healthy" mean? Is that the same as "normal"? Is there such a thing as, "chromosomally healthy, BUT…"? Does "chromosomally healthy" imply that our daughter is entirely healthy in ways that have nothing to do with chromosomes? Are there other ways to be healthy?

"Is that good?"

"Yes, congratulations."

"Then she's healthy?"

"She's chromosomally healthy."

Come on.

I'm pretty sure this is good news. I want to call Faith to find out, but she's teaching this morning and I can't reach her for another hour, so I call Tory.

"The amnio results were good...I think."

"Great!"

"The nurse said the baby is 'chromosomally healthy.' "

"Wonderful!"

"What does that mean?"

"It means shut up and relax."

Easy for her to say. Her daughter can recite the alphabet backward and count to one hundred. Still, chromosomally healthy seems like a good start, a fine place to be at four and a half months.

An hour later, Faith's response is unequivocal joy, and at last I allow myself a sigh of relief, a moment of second-trimester calm.

Maine: The Way Life's Supposed to Be

The only word I can think of to explain what has happened is *perverse*, that there has been a perversion of reality: men hijacking four airplanes and commandeering them into buildings, the collapse of the World Trade Center, thousands dead, the Pentagon, a mysterious field in Pennsylvania, the American flags hung in the windows of every house on our street, flags attached with tape to car antennas and boat masts, the photographs of the missing, poetry written on scrap paper and posted on city walls and telephone poles, written by poets as well as by people who have never before been moved to put pen to paper, blood drives, biowarfare, war. Nothing will ever be the same. By the time our child grows conscious and aware, she may not be able to conceive of a world where these kinds of things don't take place, where such things never have happened, are not even thought of in the minds of madmen. That Faith and I know of two such worlds, before and after, makes us so much older, like grandparents who sit on the porch and tell you about

World War II or the Great Depression. We are another generation, more so than ever before.

In part to get away from the awful, compelling news that draws us like moths to a flame hour after unbelievable hour, and partly to give Faith the house to herself for the weekend so that she may write Rock Opera VII, *Jesus Has Two Mommies*, I drive to Maine to spend the weekend at a friend's farmhouse.

If Boston is a million miles away from the heartbreak in New York, then rural Maine is two million. Miles of fields and pastures and rolling hills that look out across the White Mountains of New Hampshire make murder and mayhem seem like some awful movie I can't get out of my mind, like a disturbing book at home on my shelf. I'm spending a couple days with an "ex-lesbian" friend, Margie, whose mother has just undergone minor surgery back in Boston. The plan is for us to take walks through the Maine woods, visit barn sales, carve pumpkins with her husband and daughter, and then for me to drive Margie back to spend a couple days with her mother in Boston, the big, bad, awful city where lunatics are able to get on airplanes and do terrible, terrible things.

Margie was a lesbian in college and throughout most of her twenties. When her husband isn't present she still refers to herself as a "big old dyke." Despite a lifetime commitment to a sweet representative of the opposite sex, Margie ogles other women and harbors a passion for her college girlfriend the intensity of which rivals that of Katharine in *The English Patient*.

One night many years ago, when Margie was not yet a pseudoheterosexual and I was newly out of the closet and safely single, she showed up at my apartment and announced she wanted to find out whether or not we should be together.

"What are you going to do?" I was petrified.

"Spend the night."

Shit. Coming out was one thing, but having sex with someone was entirely different. At least back then. At least with Margie, who wore black lingerie and expensive coconut oil and who already had been with a number of girls.

Nothing came of our awkward, romanceless evening. All I remember is Margie saying "Your feet are like ice" and us going out the next morning with my sister for breakfast.

Winding along a trail in the Maine woods with Margie—a safe distance from her six-year-old daughter, who's blazing the way several yards ahead and out of earshot—I ask Margie why she got married in the first place, especially if she still considers herself a lesbian and fantasizes about sex with women every minute of her married life. Under the circumstances it's far more benign a conversation starter than "Do you think we should be dropping bombs on Afghanistan?"

Margie insists her choice had nothing at all to do with not disappointing her mother, wanting a child (years before purchasing anonymous sperm was as common as buying a toaster oven), or enjoying the blessing and approval of her minister father and various Presbyterian relatives. Instead she claims it was Brian, her husband, who caused her to jump ship.

"I just fell in love with him."

I generously concur that love is a many splendored thing and spare her, for the time being, my it's-hard-work-being-gay diatribe. After all, Margie's hosting me for the weekend and, more important, her daughter is only yards ahead, likely just *pretending* to be out of earshot.

"Brian's a great guy," I say.

"Of course, women are better partners."

So much for polite discretion and the possibility of an eavesdropping six-year-old.

"What's that supposed to mean?"

"Well, all of my relationships with women were so much more exciting. Women talk more, and they nurture you."

"Not all women like to communicate and nurture," I say. "Take Faith, for example. Her idea of connecting is lying on the same sofa watching the same television show as we share the same pizza. Secondly, aside from the fact that you're gay and married, I'm guessing your relationships with women were more exciting because you had them when you were in your twenties and nothing lasted long enough to involve bills, diapers, and constipation. Don't you think you're confusing gender with long-term?"

"Maybe, but I know if I had stayed with Janet, we'd be having sex all the time."

"My ass! You would not."

Margie's daughter shoots us a backwards glance.

So that's what happens. When you're gay and married you get to blame all of the negative traits of your partner and all of the difficulties of a long-term relationship on the fact that you're with the wrong gender. Just like I can imagine that if I was with a man, he would make enormous amounts of money and we would be living in stable bliss in a town with a good school system. I suppose it's no different from any coupled person comparing himself or herself to another. It's just easier to do when you're gay or bisexual. There are that many more possibilities. The grass always has more sex and money on the other side of the fence.

While I'm aware every second of every day that I am pregnant, I can't always associate that feeling with the idea of there being a person inside me, a little baby girl with half my genes. Instead it's like I consider pregnancy to be bloating, large breasts, and having to pee. I want to cathect the small life inside of me, want to imagine *her* when considering myself pregnant. A couple of weeks ago I thought I felt our small daughter moving, kicking or swimming or blowing little bubbles from a pipe, but she has since quieted down or changed position and now there is barely a rumble. (The nurse at our ob-gyn's office reassured me that this is entirely normal at twenty weeks gestation, as the baby is still very small and sensations come and go. It doesn't mean the baby has died or developed intrauterine Guillain-Barré, as I had feared.) I think because our baby fetus girl is so quiet now, on this weekend trip in Maine, far from the madding crowds both outside and inside my mind, I fall head over heels in love with Margie's seven-pound toy poodle.

I never have been a poodle person, *ever*. In fact, I was raised by my mother to be anti-poodle, especially little white toy poodles that weigh less than a cat. Poodles, my mother informed my sister and me, were neurotic extensions of their neurotic owners. They were not dogs in the true sense of the word, like a German shepherd or a rottweiler or even a springer spaniel. Poodles, like Cadillacs and satin sheets, were signs of trouble at the very least, and at worst evidence of a family's link to the Jersey underworld.

But all of that changes when I meet Sandy. Sandy is no heavier than a ball, with perpetually tearstained eyes that would make it impossible—I am told—for her to compete in any sort of dog show, and a penchant for harassing the golden Lab who lives next door to the farmhouse. Sandy runs like a stallion all day long on her little sticklike poodle legs. In the evening, when only Margie and I remain awake and are still rattling on about her sexuality, Sandy falls asleep on my stomach, her small head resting on my breasts. I'm sure it is no coincidence that Sandy happens to be roughly the size of a newborn baby. I'm certain that in all likelihood she is some safe representation of my maternal longings, which until this point have been only partially successful in their efforts to connect with the real live baby so close but still so far away inside me. To make matters even more incestuous and out of place, Sandy is my mother's name. So there you go.

Anyway, at 11 P.M., long after the sun has set into the deep, dark Maine woods and Margie and I finally have exhausted the topic of her ex-lesbianism, it's time for bed. Unfortunately, I seem to have developed an incredible and overwhelming case of the willies. Earlier in the day Margie made the mistake of telling me that slaves escaping on the freedom train once were hidden in the walls of the very room in which I am to spend the night...alone. And now the thought of sleeping in that altruistic, though most certainly haunted, room is too much for me and my imagination to bear.

After bidding my still-butch hostess good night, I stand shivering by the freedom wall. The wall intentionally borders a now-defunct fireplace, which served as a welcome source of heat on those cold frightening nights but now just looks kind of dead and creepy. I try to muster up a feeling of historical pride, or at the very least, curiosity. But the room is too dimly lit, the ceilings too low, the bed too squeaky for any form of psychic comfort. And then, holy hell! The door to the bedroom creaks open.

How frightened can one get without inducing sudden labor? The door is opening, but no one is there, not Margie or Brian or their brave six-year-old daughter, who sleeps unaccompanied in this ghost town each and every night. I stroke my stomach, trying to convince my unborn daughter that I am not afraid of ghosts and dormant fireplaces while slowly, awfully dredging up the courage to look down. At my feet I expect to find the severed hand of Harriet Tubman, maybe a young, bleeding boy crawling eternally to freedom. Instead it's Sandy the toy poodle. Sandy looks me over with her little brown eyes, and taking pity like only a little white toy poodle can, lovingly and confidently hops on the bed. I am so relieved. I know she has come to save me and the baby, to protect us until morning. Pushing back the covers with her skinny white paw and later curling up beside me, the happy recipient of all of my displaced love and nurture has done the doggy equivalent of offering me a stiff drink. In the middle of the night, when I awaken to what I am certain are African spirituals coming from the wall, Sandy gently rests her head against my stomach as if to say, *It's okay, they're free now. And so are we.*

Loading up the car for Boston, it is all I can do not to scoop Sandy up and steal her away. I have no doubt that she can protect me from all of the demons inside and out-side my head, from terrorists and war, from a fear of ghosts and chronic questions as to my own adequacy. But she belongs on this farm with this family, so instead I give her a kiss and a wink and hope to God that one day I might feel the same way about my child as I do about a friend's seven-pound poodle.

The End of the World

(OM AND A GOOD OLE BARBIE SLAM)

With all of the talk about the effect of current violent and frightening world events on children, I begin to wonder how my National Public Radio addiction might be affecting our unborn child. Twelve hours a day of world news delivered with that carefully modulated, folksy-seriousness of all NPR commentators could serve to create either an overly aware anxiety-ridden child or a toddler bizarrely attached to the voices of Bob Edwards, Noah Adams, and Scott Simon. Some days our daughter hears their voices more than she hears Faith's, certainly more than she hears the voices of her grandparents and aunts and uncles. It's not that she understands what they are saying, what "World Trade Center," "Islam," and "biological warfare" mean. But surely she senses their tone of urgency, the feeling of fear and drama in their voices, not to mention the fear and drama it elicits in me, her vessel. I decide to try to kick the NPR habit, if only by a few hours a day, and provide her with some other form of auditory stimulation: jazz, reggae, maybe even silence.

At prenatal yoga each week we are accompanied in our various poses by a tape of monks chanting "om" over and over again. I decide it would be far better for our daughter to hear a gentle "om" than *Talk of the Nation* and special reports from John Ashcroft. I think I might like the chanting monk tape to serenade my labor, but am afraid to tell Faith. If ever there was something I knew in advance we never would agree upon, it's monks chanting in the delivery room. While Faith has promised to join me at a special prenatal partner yoga session, I just can't see her opening up our birth experience to the sound of grown men and women droning on and on.

I suggest to Faith that we begin thinking of music we might want to bring to the hospital.

"What if it's the Super Bowl?"

"The Super Bowl?"

"Your due date is early February. It could be Super Bowl Sunday."

"You've got to be kidding."

"Wouldn't you want to see it?"

"I'd be in labor!"

So much for monks chanting.

Back on the West Coast, Kevin, my sister's husband, has been working for weeks burning CDs for their delivery. "It's got the greatest music on it," Carrie tells me.

"Faith wants to watch the Super Bowl while I'm delivering."

"Well, you like football. And we're bringing a magnetic backgammon board. After the epidural," Carrie explains, "you have a chunk of time when there's nothing much to do."

I don't want to burst her bubble and remind her that by

that point she'll likely have been awake for twenty hours and might, in fact, prefer to use the time to sleep or cry or try to eat something.

My sister is getting very close to her due date. She's walking around Los Angeles two centimeters dilated with various bodily liquids dripping down her legs. In addition, she is carrying very, very low and basically has a head pressing against her cervix all day and night, making her feel that the baby could drop out of her at any moment—at an audition, in the car while she's stuck in traffic, at the Beverly Center. So if the fantasy that she and Kevin will be playing a few rounds of backgammon in between contractions helps her maneuver from day to day, minute to anxious minute, so be it.

"That's nice," I told her. "I want to bring a tape of monks chanting 'om.' "

"Do you think that's good for the baby?"

"It's better than NPR."

"What does NPR have to do with it?"

Winding our way through the hilly neighborhoods that lead to the arboretum, Faith and I pass four young children, each no more than ten years old, playing in front of a large gray Victorian house. Each child has a naked Barbie doll with a long string tied around its neck. The kids are swinging the Barbies by the strings and smacking them against the sidewalk over and over again. They giggle and laugh and egg each other on as Barbie after Barbie collides with concrete.

"Are you guys having a Barbie Slam?" Faith asks casually, as if Barbie slamming—naked Barbie slamming—were an ordinary playground sport, right up there with tetherball and capture the flag.

"It's Destroy-a-Barbie," a girl with shoulder-length brown hair and a devilish grin remarks.

"Oh," Faith says.

An hour later, on our way back from a walk in the arboretum, the children are gone but the Barbies remain, dismembered though they are, their plastic body parts lined up in neat rows on a wet square of cement. It looks as if the Barbies have been hosed down, cleaned off, and laid to rest with the precision of either a mortician or an obsessive-compulsive sociopath. The whole scene is frighteningly reminiscent of our daily news reports, in which the images of bodies falling from buildings to their crushing death on the streets and sidewalks below are suggested but not shown by disbelieving anchormen and women who leave the gory details to our—and our children's—imagination. I wonder if these children have been watching the television news behind their parents' backs, or if their parents simply leave the television on all day long for anyone to see and hear.

Faith thinks I'm reading too much into an innocent game of Barbie Slam or Destroy-a-Barbie, whatever it is, and that, of course, causes me incredible doubt as to our ability to raise a child together.

"Naked Barbie parts on the sidewalk? How could that not be a child's expression of what just happened in New York?" I ask her.

"Kids always hurt Barbies."

"They do?"

"Sure."

"Did you ever torture and then dismember one of your Barbies?"

"No."

"So."

"So?"

Lately it seems Faith and I have been locking horns; the most opinionated and uncompromisable parts of our personalities have been leaking out and grating against one another like nails on a blackboard. This time it's "Harlie the overanalytical versus Faith the realist." Who will win? Whose personality will dominate our household and raise our child? It's a personality slam, the two of us swinging our personalities around by strings and slamming them into each other over and over again.

When we bought the house it was a version of the same conflict: Now that we own a home, what kind of home is it going to be, the boundaryless communal space of Faith's dreams or the private, boundary-filled refuge of Harlie's? Back then we had to resort to couples therapy to regain our footing. Once a week for six months we met with a social worker in the basement office of a rambling Victorian not unlike the Barbie Slam house and discussed how someone who preferred that the upstairs lesbians not possess the key to her apartment could share a life with someone who had just invited said lesbians to let themselves in whenever they needed. We stopped fighting, the sessions working not so much because the counseling was effective as because Faith and I were able to reestablish our love for each other every week by bashing the therapist's choice of office decor.

We would arrive at her office all huffy in our separate cars and then sit in the waiting room hysterically laughing at the watermelon toilet seat and watermelon face towels in the bathroom. When our appointment time was switched so that our session immediately followed group therapy for obese women, Faith and I were lovers again, tried and true. Nothing against obese women, or our therapist, who also was obese. It's just that since Faith and I are not, the whole thing made us feel really strange and small and very likely the object of disdain by our own therapist as well as the women in the group ("Whatever are the pint-size lezzies going to whine about this week?"). Because we kibitzed about all of this ad nauseam, we were soon communicating just fine, thank you very much, and no longer in need of counseling. Or so we thought.

Those Deadly Maine Roads

There is sad news waiting for me when I get home this evening. Sandy, the toy poodle, has died. She was hit by a car on the beautiful but treacherous rural Maine road that winds past the farmhouse. So much for Maine. So much for escaping the random horrors of the rest of the world. Sandy was killed instantly. Margie saw the whole terrible thing and is devastated. It's almost too difficult for me to imagine, to have such a small and innocent creature in your arms one day and then gone five days later, and to see it happen. I tell Margie that Sandy was happy in the last days of her life, chasing the golden Lab and running through the fields. I tell her being killed instantly is better than so many other passages into the next dimension.

It's 11:30 P.M.; I've gotten in late from baby-sitting Simone. Faith stayed up to give me the news. Now she pats my back as I sit all sad and droopy on the edge of the bed.

"I know how you felt about that dog," she says, trying her hardest to empathize.

Later, lying next to Faith, hands on my belly, I allow

myself to consider the unthinkable: What if this had been a child? What if this had been our child? Having a child means living with the potential for earth-shattering loss. I suppose loving anyone means living with the potential for earth-shattering loss. This is just the beginning, the feeling I had for Sandy. The transient awareness of there being a baby inside me, who one day will be out in the terrifying and arbitrary world, is just the beginning of looking beyond myself and loving with all my heart another who I will not ultimately be able to save or even keep one hundred percent safe. It's suddenly impossible to believe that people who lose their children go on living, that they ache forever and stay up at night crying in the dark but still can feed themselves, go to a movie, take a walk. How is it that we have gone on parenting under these conditions? How is it that we continue to reproduce and love unconditionally in a world of hijacked airplanes and cancer and speeding Fords?

I say a prayer for sweet Sandy, who rested her small head on my breast and prepared me to love something so small and fragile. I wish her spirit peace and rest and then close my eyes.

Daddies and Faith

(A Baby Sister's Baby and Pussy-a-Go-Go)

It's 10 A.M. Saturday and we're at prenatal yoga for partners when suddenly it dawns on me that Faith is a woman
and we both are gay. I really hadn't thought too much
about this seemingly overt fact since the days of regular
intrauterine inseminations, when it was glaringly obvious I
was not having sex with a man. It just hasn't been that big
of an issue, not at our prenatal appointments, not at childbirth class, not even at infant CPR. It's really not until now,
at a well-meaning prenatal yoga for partners workshop,
that it again becomes acutely apparent.

There are eight couples altogether, including us, four on
each side of the room. It's not the fact that each of the other
couples consists of a man and a woman that drives the point
home. And it's not that, in an effort to be inclusive and polit-
ically correct, our yoga instructor compulsively refers to each
couple as "birth mother and her partner." It's not even that
when we have to hug and caress each other openly and in
public, I catch a couple of the other "partners" sneaking a

peek at the two gals, free soft porn on a Saturday morning. It's that Faith can't hold me up.

In all of the positions that involve "partners" propping up the birth mother, Faith is either too short or too weak to support me and my pregnant body. While the other birth mothers safely sink into the large, hairy arms of their male partners, more than once the weight of me causes Faith to lose her footing. These are labor support positions, poses that we are supposed to practice and make second nature so on that day of days, when I am experiencing contractions so strong and painful I want to gouge my eyes out with a fork, I instead can transfer all of my weight onto Faith and focus on breathing deep, meditative breaths. So much for labor support. To make matters worse, Faith is an inch shorter than me. Leaning into her means leaning *on* her.

Our yoga instructor has the monk tape playing, and a peaceful "om" fills the room. The other couples sway rhythmically to the tune of their socially sanctioned and physically coordinated love. Each couple is instructed to breathe together, to move together, to open up their psyches to the baby within and receive each other's love.

I try hard to believe I can drop backward without looking and Faith will receive me with open arms. This time it's not that I'm a suspicious and inherently distrustful person that prevents me from believing in the power of my girlfriend to catch and embrace me. It's that she can't. I've put on twenty-five pounds in the past six and a half months. Each time I give Faith my fertile body to hold she makes a grunting sound in my ear.

"Stop acting like I'm killing you."

"You are."

The yoga instructor calmly works the room, stopping by each birth mother and partner to adjust and approve, all with

the slightest touch of her hands. When she gets to us she does-
n't know what to do. All those years of yoga training, vegetarianism, and spiritual healing haven't prepared her for a 140-
pound woman pressing into a 112-pound woman who is supposed to be holding her so securely as to make her feel weight-
less. I suppose the same might be happening if I had gotten
involved with a very small man, or a very weak man, or a man
who was physically challenged in some way. It's not necessarily gender-specific, the fact that, while the other birth mothers
get to lean back and rest their heads on their partners' shoulders, when I stand in front of Faith she actually disappears.

"Try this," says the yoga instructor as she carefully
bends my knees, arches my back, compresses my shoulders.

It's like I'm lying back in a dental chair but without the
chair, and with Faith's chin pressing into my head. It's the
most uncomfortable and precarious position I can imagine.
If a breeze blows in through the open window we're both in
serious danger of toppling over. At least I'm not really at the
dentist's, I tell myself. Be grateful for small pleasures.

"There," says the yoga instructor, then moves on to the
next couple.

Faith grunts again.

"Do you have me?" I ask, thinking that maybe, somehow, this is what the position is supposed to feel like,
maybe it's some yogic interpretation of life and death, grav-
ity and a lack thereof.

"Just don't move."

Is this our destiny for labor and delivery? Two small
women alone doing a job meant for a man and a woman.

"It's not like we were really going to do those positions anyway," I tell Faith later. "I mean, I'll probably be
in so much pain I'll be yelling at you to get the fuck away

from me. Trust me, we won't even remember them."

Secretly, I wonder if men and women remember them, if part and parcel with putting a penis inside a vagina in order to make a baby comes the uncanny ability to remember all of the labor support positions you've been taught and to perform pre-natal partner yoga. But maybe that's just me again wondering whether or not we've done the right evolutionary thing by bringing a child into this world, wondering whether the two of us really can manage so vast and incredible an undertaking.

"I'll remember them," Faith says without batting an eye.

I'm so relieved. Of course we'll be okay! Of course we're doing the right thing! Didn't we breathe in beautiful syn-chrony with each other during warrior pose? Didn't we relax deeply and without laughing during the twenty minute med-itation that concluded the workshop? Haven't we been together for close to ten years already and learned how to make our love for each other flourish and grow in a PIV (penis in vagina) world? We can do it. We must do it.

To prove her point, Faith pulls me back into her and digs her chin into the top of my head. I oblige by bending my knees, slouching my shoulders, and letting go all doubt and disbelief.

At 10:15 on an unseasonably warm October morning, my sister calls to tell me she's having contractions. "They're five minutes apart," she says. "But they don't hurt."

"Did you call the doctor?"

"No."

"Call the doctor." Why she and Kevin haven't already come up with this idea is beyond me.

"But they don't hurt."

"They will."

Two hours later Carrie and Kevin are walking the halls

of Cedars-Sinai hoping to encourage Carrie's body to go into active labor. Their doctor says the baby will be born within the next twenty-four hours. It's unfathomable how much there is still to go through until then.

Two more hours pass, and they call to say the doctor has broken Carrie's water with a knitting needle. My sister is scared. And suddenly it seems really, really wrong that they are there and we are here, that my sister is about to deliver a baby in Los Angeles and Faith and I and my mother are in Boston. It had been my sister's idea for us all to visit her in shifts after the baby was born. My mother was instructed to arrive first, immediately after Carrie got home from the hospital. Faith and I were told to come out once my mother and all of Kevin's various parents and stepparents had left. Carrie sealed the request by telling us she wished people would just ask her what she wanted rather than announce when they were going to arrive for a visit. So, "Okay," we said. "We'll come in shifts."

"I wish you were here," Carrie is crying now over the phone.

All I can think to say is, "We know for next time. Next time we'll come as soon as you think you're in labor."

"Next time?" Apparently the idea of going through this again is not exactly what she wants to hear. "Call me again," she says, and hangs up the phone.

Our next conversation is even more upsetting. They gave Carrie the contraction-worsening Pitocin that she had been fearing from the moment she found out she was pregnant. "A huge dose," she says, her voice trembling.

"How bad is it?"

"I can't even tell you," she whispers, a soldier back temporarily from the front lines thanks to a recent epidural. "I just had the epidural. It's warm. It's working. I think there's relief."

She sounds traumatized, maybe even crazy.

"Relief." She repeats the word over and over again like a prayer and then quickly says, "Call me back."

I don't exactly know why she keeps ending our conversations and instructing me to call her back. I can only guess it's all she can do to say a few sentences at a time. I hang up and wait what feels an appropriate amount of time for her to regain her strength and then call again.

"How are you now?" I know better than to ask if she and Kevin have played any rounds of backgammon.

"I had to have another epidural," she weeps, devastated, betrayed by the sick joke of the first one's inadequacy.

"Shit."

"I love you."

My sister and I, despite loving each other very much, never say "I love you." We just didn't grow up in an "I love you" kind of family, the kind where "I love you" is said at the end of every phone conversation, much like "goodbye" and "Don't forget to turn off the light in the garage." Carrie's either seriously terrified or incredibly high.

"I love you too," I say. It's insane that Faith and I aren't there, totally and utterly insane.

I want to talk to Kevin to find out more, but Carrie won't give up the phone.

"Call me again," she says, and hangs up.

There seems to be something hugely and immensely wrong with evolution for human childbirth to still be such hell on wheels. Nature has had thousands upon thousands of years to evolve the method, and still women are brought to the brink of hysteria by the pain of it. Someone once told me the difficulty is that the human head (i.e., our brains) has grown so big over the years while women's cervixes and vaginas have not compensated for the growth (thank goodness).

Whatever, it seems a deeply flawed system, so deeply flawed, in fact, that it causes me to doubt the entire theory of evolution, even Darwin himself.

I make what will turn out to be the final call to my sister as a "nonparent."

"How are you?"

"I'm nine centimeters dilated. They're going to make me push."

"That's great."

"I'm scared."

"The worst is over. You'll do fine. Just push that baby right out." I'm talking out of my ass. I have no understanding of any of this, just that you're supposed to be encouraging during labor, to say positive, supportive things. "Pushing is a piece of cake. You'll have that baby in your arms in no time."

"Think so?"

"Sure."

There is a small rumble in my lower belly, a little girl rolling around in confirmation, a fetus who—like her cousin—one day soon will want to come out and become a baby. I begin to picture Faith and me going through labor together, and then my brain shuts down.

There's nothing more to do for my sister but to wish her and Kevin the best of all possible luck. Then Faith and I light a candle in honor of them and their baby and the birth, and sit back and wait for the phone to ring.

Less than two hours later, Carrie is holding in her arms a healthy, beautiful golden-haired boy whom she and Kevin name Jonah.

Our daughter is a gymnast, a kick-boxer. She is a modern dancer, a very limber yoga devotee. Not a day goes by now when I don't get a wallop to the intestine, a kick to the ribs, a squeeze of my gall bladder. It's good for me, being reminded by this little girl that she's here and has a life already, a little fetal Daytimer she uses to plan her activities: 1 A.M.: Poke Mommy in the bladder; 3:45 P.M.: Try to kick through skin. And all day long I feed her and water her and vitamin her. It's like we're a team. And today, after twenty years of vegetarianism, the team is going to have a corned beef sandwich for lunch. Go team!

I make a mistake. One night when I can't sleep I reread our donor's profile. At six months gestation, with our little girl punching and rolling around inside me, we—at least *I*—have become increasingly detached from our donor. It's like he's over and done with for the time being. The male half of this baby seems more representative of all of the men in our life whom we love and feel close to rather than some strange guy and his strange family. But while looking at the donor profile at first seems like a fun thing to do, before long it totally freaks me out. It's like throwing on fluorescent lights in a room once dimly lit by candles and the glow of burning incense: Oh, my God, I am carrying a stranger's child! I have blended my genes with those of a total stranger! Not only a stranger, but the kind of stranger who would sell his sperm for money.

The baby does a half-gainer with a twist. Who is she? What will she look like? What have I done? What in God's name were we thinking? That I have been impregnated by Faith? By love? By wishful thinking? And then there are this man's siblings and parents and grandparents, and holy shit, I file the profile away again.

There is a time and place for denial, for locking the truth

up in a file cabinet during the period in which our little girl is born and looking in our eyes, during the time when it matters not a bit from where or whence she came. One day she will want to know. Until then we will live lives of blissful oblivion, believing our child is just some version of me and Faith rather than the product of me and some dude in California I might not even want to have coffee with. Thankfully, it's all enough to put a pregnant girl back to sleep.

My mother has gone and returned safely from Los Angeles. She has held her first grandchild. She has watched her youngest daughter become a mother. She has not had to see a doctor in two months. Though she is quite thin and of minimal energy, we are close to normal once again, a family with a relatively healthy mother who travels by airplane and doesn't have to visit the Dana-Farber Cancer Institute every week. There is a baby and a pregnancy and so many sighs of relief. Until it is our turn to visit Carrie, Kevin, and Jonah in Los Angeles.

Carrie bumps our visit back a few weeks to accommodate the slow, endless trickle of stepparents that has crept ever so insidiously into her life. It's like there are stepparents crawling out of the woodwork, various exes and newly married second husbands suddenly darkening my sister's doorway and claiming rights to a grandchild as close in blood as a chimpanzee is to a giraffe.

With each week Carrie bumps back our visit I gain like ten more pounds, and I am no longer able to sit comfortably for longer than thirty minutes at a time. My face is as swollen as a cantaloupe. My post-9/11 anxiety/pregnancy-nesting quotient has quadrupled.

"I can't wait to see you," Carrie says. "But we have to

wait another week because Kevin's father's second wife decided to stay longer."

She sounds weary, tired to the point of total acceptance.

"I think I'm running out of time."

"What do you mean?"

"Soon my ass won't fit into a coach seat."

And then there's our mother's two-month appointment.

"Stay for the appointment," Carrie insists. "We'll come out there."

"When?"

"January."

"But I'm having a baby in February. Come in February."

"We'll come in March."

"Then I won't see Jonah until he's five months old."

"Then come out here."

"You just told me not to because of all of Kevin's parents!"

"Then come in December."

"I'll be too pregnant."

"Then I'll see you in January!"

Sometimes I have the fantasy that if Faith and I lived in Los Angeles, we all would be bound together in sisterly love: her sister in Silver Lake, mine in Marina del Rey, us somewhere in the middle. And other times I wonder.

The kitty and I are having a heck of a time. It's like he knows what's coming, and rather than having the maturity to support his caretakers and respect their decision to have a baby, he's behaving like a cat. When I find him sleeping on the changing table or in the bassinet and kindly ask him to move, he digs his claws so deep in whatever it is he's lying on—an afghan knit by my grandmother, the newly cleaned bassinet mattress—that lifting him up and out means dragging every other item along with him. He cries constantly to be petted, and when I pet him he tries to bite me. My mother is so distressed at the thought of there being a cat in the same home as a newborn baby, especially her newborn grandchild, that she confesses to having posted a sign at a nearby animal hospital: FIFTEEN-YEAR-OLD CAT NEEDS A HOME.

And then the secretary at our research lab announces one day, "I'm just waiting to hear that cat of yours has dropped dead."

Even Faith admits that living with this particular kitty is like living with a really strange roommate.

It's true the kitty has some unusual habits, like meowing incessantly, leaving balls of poop in odd places around the house like on top of the piano, and sitting by the bathtub and crying until someone takes a shower so that he can lick water droplets from the wall rather than resort to his water bowl. He's destroyed every piece of furniture that isn't made of solid wood, vomits only on carpet or clean laundry, and has woken me every night at 3 A.M. for the last twelve years by meowing loudly in my right ear. Still, he's a helpless kitty and I promised to take care of him twelve years ago when a good friend decided she no longer could keep him. (It was too much to worry about him and her newborn baby!) So I am devoted, or stuck, as someone else might say. And after all, hasn't he taught us how to keep a small creature alive and well?

Now in five days our strange roommate turns sixteen. It would be too awful to admit that I actually have entertained the thought, *How much longer?* If I did, no one would believe that I'm an animal lover, that I grew up with dogs and cats and guinea pigs and goldfish. It wouldn't make sense, an animal lover wondering when her cat will die. So I won't admit it to anyone, ever.

Instead I immediately race to the animal hospital and take down my mother's sign. Rather than banish the kitty, we decide to try and teach the old boy some new tricks. We start by banning him from our bedroom when we sleep. If these are the last two and a half months of my life to sleep through the night, then by God I'm going to do it.

It's hard. Though rude and feral at 3 A.M., the kitty is soft and cuddly at 11 P.M. when we go to sleep. Still, we are learning to be parents. We are setting limits and making rules. So out he goes.

Next we have his claws cut short. I would do it myself if I could. I would do it every week, give him a little kitty manicure. I would play videos of cats or birds and wedge cotton between his toes before soaking and filing down his deathly daggers. Unfortunately, clipping the cat's nails is comparable to learning to fly a plane: It's dangerous and difficult and just not going to happen in this lifetime, not by me anyway. So it's off to the vet. By the time the baby gets here he'll have stubs for claws and will have learned to stay out of our bedroom.

At breakfast the day after the kitty's manicure, Faith reads aloud an article that claims researchers have found that children raised with animals are less likely to develop allergies to them later in life. The kitty meows an exasperated little meow as he futilely attempts to shred a rattan chair.

"Tell your mother," Faith says to him.

"Meow," repeats the kitty.

"All right," I grumble.

"Meow!"

"I said all right."

"Meow meow meow."

The Third Trimester, Or, The Party's Over

(DIVORCE, CELL PHONES, AND MARY CHAPIN MAGDALENE)

The world's easiest pregnancy is getting a bit uncomfortable, just as predicted. A searing pain under my right breast comes and goes with no apparent regularity. At first I think it has to do with what I'm eating, that dairy products are the culprit, or fatty foods. I lighten up on the cheese, the pizza, the fries. It helps for a while, and then there it is again, a small fire blazing below my ribs. I load up on Tums, try Zantac. Nothing works. One day, in exasperation, I simply lie down. The pain recedes. I try turning to my right: holy hell. To my left: relief. Walking, relief. Sitting, agony. The answer: moving. The problem: the position of our little girl. She has settled already with her head down, comfortably stretching her feet all the way to my right breast, where she likes to kick and turn and squeeze her tiny toes. For this to have occurred, our daughter, living from pubic bone to rib cage, my uterus, it seems, must have expanded to almost the size of my entire torso. That freaks me out even more than the fact that there is a person living inside me, kicking the life out of my gall bladder, that an organ previously the size of a fist has inflated into a mammoth balloon. It seems entirely possible that if someone were to turn me the wrong way, or if there were a sudden change in atmospheric pressure, I might be sucked up inside myself, like my uterus could take over my body, maybe even the entire earth. And then one day the pain is gone. That's that. My small body-mate just up and migrated farther south for the remainder of her stay.

Two weeks later the next unexplained symptom of pregnancy appears. Every morning, about five minutes after finishing breakfast, I come close to passing out. I'm fine until I eat, and then all of a sudden I get short of breath, lightheaded, a bit nauseous, and my resting pulse soars to over a hundred. One day at work I have to hang my head between my legs for five minutes in order to steady myself enough to be able to walk down the hall and ask a doctor what's wrong with me.

Unfortunately, I work in a psychiatric hospital, and psychiatrists, though the least informed about medical issues such as fainting pregnant women, are the most likely to offer medical advice, even if they have no idea what's going on. While internists hate being asked medical questions by their friends and neighbors, psychiatrists *love* it. It makes them feel like real doctors. Within moments after my workplace orthostatic hypotension I have psychiatrists lining up outside my office door to offer their opinions.

"You ate too much."

"You didn't eat enough."

"The baby is resting on your vasovagus nerve."

"You're anemic."

"You're anxious."

Without hurting their feelings or insulting their medical egos, I excuse myself and call my obstetrician's office.

I am seven and a half months pregnant and having quasi–fainting spells every morning when Faith and I have perhaps the worst knock-down drag-out fight of our entire relationship. It begins on a Sunday, but Faith doesn't find out about it until Monday evening, because she's in Virginia and I'm home in Boston stewing and brewing myself into a frenzy. The issue at hand—connection, communication, paying attention. Sounds simple. There is no affair or fling or sexual betrayal. There is no name-calling or knife-throwing or nervous breakdown. No one has lost the family riches at the races or insulted the other's mother. There is no deep, dark secret, no midlife crisis. There is only this: Faith did not call home for thirty-six hours. And when you are seven and a half months pregnant and passing out from Raisin Bran, that is cause for war: full-scale, take no survivors, land, sea, and air war. You'll see. *Just wait.*

The catalyst for this war is Mary Chapin Carpenter. In addition to having a beautiful voice, Ms. Carpenter is very kind, so kind in fact that she agreed to film a cameo for Faith's new rock opera. Mary Chapin Carpenter kindly agreed to play the whorish Mary Chapin Magdalene in *Jesus Has Two Mommies,* on the condition that Faith and the crew drive to her Virginia farm to film the segment.

The decision of whether or not Faith should go was easy: Of course she should. It's not like I am bedridden, we tell each other. It's not like we have a newborn baby! What we don't say, of course, is that Faith should either phone home regularly or leave a way for me to reach her while she's on the road. Who needs to utter those words aloud? That would be like reminding her to go to the bathroom. You don't need to tell your girlfriend to call you when you're pregnant to make sure that you and the baby are okay, that you haven't gone into premature labor and are lying alone in an emergency room, that the baby isn't incubated in a neonatal ICU. Because she'll *want* to call. Since the searing rib pain and orthostatic hypotension of the third trimester have kicked in, we've checked in with each other several times a day. So who would even think of bringing it up?

Sunday morning at dawn Faith leaves with Bill, her producer, and Eric, the guitarist who stayed too long. We shout our goodbyes from bedroom to hallway and back again.

"Have a great time! I can't wait to hear all about it."

"I love you guys [me and the baby] so much!"

"We love you!"

And then I don't hear from Faith until Monday. As in the next day. At 4 P.M. As in thirty-three hours later. Monday. That's when Faith asks Mary Chapin Carpenter if she can borrow her phone to check in with her girlfriend. That's when, on the phone in Mary Chapin Carpenter's beautiful farmhouse, Faith gets the surprise of her life: me screaming bloody murder like I haven't done since age thirteen when my father announced he was moving our entire family from New Jersey to Massachusetts. And then I slam down the phone.

Faith reports later that she was in utterly unanticipated shock. What did she do wrong? On Sunday they had driven for twelve hours into the netherlands south of New England, where everyone's cell phones suddenly stopped working. Late, late, late, long after a pregnant woman would have gone to sleep, they found themselves bedding down for the night at a motel without phones. The next morning they raced to the farmhouse to begin filming. As soon as the Mary Chapin Magdalene video was complete, Faith swallowed her pride and asked to borrow Ms. Chapin's home phone. Unfortunately for Faith, this story does absolutely nothing to alleviate a pregnant woman's hurt and rage.

I didn't start off mad. Late Sunday afternoon, after Faith and I had been apart for twelve hours, I was fine and calm and left understanding messages for her on her cell phone. The first one was friendly enough: "Hi! It looks like your cell phone's not working. Why don't you leave a message at home when you get a chance and let me know how you are and how I can reach you."

Eventually I got the idea of leaving messages on our home voice mail.

"Hey! Let's leave each other messages here!" I said, amazed at my own ingenuity. "That way you can check my messages and I can check your messages and we can be in communication until your cell phone works again."

Only Faith never checked the messages.

"Hello, Faith," was my third message. "It seems I have no way of reaching you, since your cell phone doesn't work and you're not checking messages. I've never known you to not check messages, but I guess for some reason you're not."

The sixth message went something like, "So, let's see, I'm seven and a half months pregnant and you don't feel the need to see how I'm doing or for me to be able to contact you. This feels really irresponsible and dangerous."

To make matters worse, because Faith isn't checking our voice mail, whenever I check to see if she has called in, I have to listen to my own furious and hysterical messages, and that, of course, makes me even more furious and hysterical.

By the time we speak directly on Monday afternoon I've broken up with Faith in mind, sold our half of our two-family house, and moved in with my mother. Or, somehow, I've bought out Faith's half of our half and continue to live there to raise our daughter as a single parent, renting out a room or going on welfare to pay Faith's half of the bills. In a more self-aggrandizing version, I suddenly inherit enough money to buy Faith out, live as a single parent, and not have to go back to work or go on welfare. Sometimes in addition to inheriting money, I suddenly earn tons of money from my writing, and then I raise a child alone as a famous and wealthy author. Whatever the case, Faith is nowhere in the picture, erased with emotional white-out for not caring enough to keep in contact despite distraction by a very kind Grammy award winner.

And then Faith calls. "Hi, we're still in Virginia..."

After having been rendered helpless and incommunicado for the last day and a half, slamming down the phone feels like my only emotional recourse, the only perverse form of control I have left. SLAM! It's initially satisfying but soon after feels embarrassing and melodramatic. Still, I do it again and again. Each time Faith calls from the road (for some reason, on the way back she's able to find and use telephones) to find out what in heck is going on, I blurt out a wild and crazed version of my hurt and pain and then slam down the phone. Only I don't really "slam" the phone down. I press the reset button. It's a silent form of slamming down the phone, a modification made for the sake of our unborn child, whose budding auditory system I don't want to damage by actually slamming down the receiver. Pressing the reset button reassures me that, although I'm enraged and out of control, I'm still being a good parent—which certainly is more than *some* of us can say.

"So let me ask you something," I pant during one such conversation, so upset and pregnant that I quickly become short of breath. "During the entire twelve-hour trip, did you once stop to go to the bathroom or get something to eat?"

"Yes."

"And were there pay phones at any of these places?"

"There were long lines."

"Were there long lines to buy food?"

Silence.

"Because I'm sure the long lines didn't stop you from waiting for a CHEESEBURGER AND FRIES."

SLAM/PRESS.

"You could have called from the road, from the front desk at the motel, from wherever it was you had breakfast the next morning, from Mary Chapin Carpenter's living room *before* filming."

SLAM/PRESS.

I deliver my argument like a jackhammer, pounding it out again and again. The point I'm trying to get across, the one from which I'm hanging by the collar, is that Faith could have called if she had wanted to. But for some reason that terrifies me, she didn't want to. In my new single-parent mind, it's as simple and horrible as that.

I'm so upset I tell Faith not to come home. She's due back around 3 or 4 in the morning, and I don't want to be lying awake in bed drowning in rage until she walks in the door. Because then, even though it'll be the middle of the night, we'll continue fighting, and I am very pregnant and need my rest and the child inside me needs rest and for me to be calm. So I tell Faith not to come home. It's the first time in eight and a half years either of us has said this to the other, including the time Faith confessed to kissing another girl. She is devastated.

"You're just trying to hurt me like you feel I hurt you." So what. "I need to get some sleep."

It's not until days later, after Faith and I are sleeping in the same bed again, when I hear her refer to the restaurant at which she, Bill, and Eric had dinner prior to their stay at the motel, that I insist we again enter into couples therapy.

"You had dinner at a Denny's?! They have pay phones!"

Cancer Lessons

The major lesson from long-term relationships, from pregnancy, from illness, is that nothing ever goes as planned. Our worst fears are never as terrible as we have anticipated, our most glorious accomplishments never as life-altering and unequivocal. What looks like a great day coming down the road can turn out to suck to high noon. A day with all of the earmarks of hell on earth can, in fact, be filled with splendor and light.

In the course of my mother's illness she has gone to the doctor feeling healthy and full of energy only to be told there is cancer lurking in places we hadn't even thought to worry about. Similarly, she has gone to the doctor emaciated and in need of a wheelchair only to be told that the cancer has not grown and she is free for two months to eat and play and catch her breath.

One of the moments I've feared most during the course of my mother's illness has been that trip to the hospital when the doctor puts down her pen, closes my mother's hospital records, and says, "There's nothing more we can do." But just like all other enormous moments, both good and bad, that too will not go as planned. It will be neither better nor worse, just different.

At the end of my mother's two-month medical hiatus, sure enough she returns to the doctor on her own wind— no cane, no wheelchair—only to find the cancer has resumed its dreaded advance, this time rendering her no longer eligible for chemoembolizations. The liver doctor refers my mother back to her original oncologist—a kind and gentle woman whom my sister and I spotted in Provincetown two months ago holding hands with another woman and laughing as our mother lay listless and recuperating on a sofa in our rented cottage.

The appointment with the presumably gay oncologist is scheduled for next Tuesday. Because I'm currently suffering from my own bodily conundrum, I'm unable to accompany my mother, unnerved and trembling as I am at the office of my ob-gyn. Therefore much of what I learn about my mother's condition comes from a phone call I make to her doctor days later, after my mother offhandedly mentions that she has started systemic chemo again, that the kind and gentle lesbian oncologist walked her to the infusion room herself so my mother might begin treatment immediately.

"What. Why?" I have so many questions, too many questions. "What's the name of this chemotherapy?" I ask my mom. "How long do you have to have it?"

"I don't know."

"Well, what are the side effects?"

"Not bad."

"Did you ask?"

"It may not make my hair fall out."

"Anything else?"

"I didn't ask."

It really is too much for any soul, especially one riddled with cancer and traumatized and exhausted by five and a half years of chemotherapy, to attend a doctor's appointment alone and be in her right mind enough to ask pertinent questions and retain information. I remind myself that my mother is skinny and afraid and doing the best she can. I remind myself that I have questions no mortal can answer. Like, Why is this happening to my mother? Why is this happening at all?

Talcum powder. It all boils down to talcum powder, my mother's only risk factor for developing ovarian cancer: repeated long-term talcum powder use in her underwear, in her diaphragm. Some malignant ovarian tumors have even been found to have talcum powder inside them. I want to sue Johnson & Johnson. I want to paste stickers on tollbooths across the nation that read TALCUM POWER = DEATH.

Instead, "Can I call the doctor and ask her a few things?" I ask my mother, figuring this is the next best thing to accompanying her to the doctor.

"No," my mother tells me.

"I can't?"

"I'll ask her the next time I see her."

I decide to make the call anyway, against her wishes. Defying my ailing mother seems like the right thing to do. Her "don't ask, don't tell" policy no longer seems in good judgment. Besides, I think I know why she doesn't want me to call: I think the idea of her daughter calling any other adult to discuss her well-being makes her feel demeaned and belittled, like an invalid, and that is just not a good enough reason.

I defy her wishes one day after noticing that she's thinner than ever and hardly cares about eating; that she sits down while shopping, instructing a salesgirl to find and bring to her elastic-waist pants in any color but brown; that there are dark rings beneath her eyes; that she stoops like my grandmother did at ninety; that her nose is perpetually running and her back is always hurting. I defy my mother because I need to know the truth about her condition, even if she's not ready to know. I call the gentle lesbian doctor to find out if there really are a "bunch of options" as my mother has told me. I call convinced the only reason my mother doesn't want me to call is so I don't insult her and her ability to take care of herself.

The doctor calls back two hours after I leave a message with her secretary, and suddenly I'm a different kind of daughter, a different kind of person. Suddenly I have a handful of reality not bathed in my mother's hope and reassurance. So this is why she didn't want me to call. She wanted to protect me, to spare me the truth that somewhere deep inside, despite her perpetual air of optimism and denial, she knows quite well.

The doctor confirms that the reason my mother has been referred back to her by the liver specialist is that the cancer is spreading. She tells me the reason she has chosen this new chemotherapy isn't because it will not make my mother's hair fall out but because it is the only one she feels my mother can tolerate right now. My mother's liver is in pretty bad shape and can barely metabolize a compound as toxic as chemotherapy. Therefore she's being treated with this chemotherapy because it's cleared by the kidneys instead of the liver. There's a twenty percent chance it will be effective in stopping the cancer's growth, never mind shrinking it, never mind putting it into remission.

"The cancer seems to have the upper hand these days," I say, hoping the doctor will disagree, hoping she'll say something upbeat and optimistic just like my mother would.

But "Yes," she says. "Yes, it does."

Someone once told Faith and me that the first week home with a new baby is "sheer and utter hell," the next three months just "sheer hell," and then it's all bliss. That person doesn't know what she's talking about. Sheer and utter hell is your mother battling cancer for five and a half years, deteriorating before your eyes, failing while you're about to give her a grandchild, knowing that when she talks to your stomach she's trying to imprint her voice on her tiny granddaughter's mind and heart for all the years she won't be there to talk to and hold her. That is sheer and utter hell, not life itself. Sleepless, diaper-filled nights are evidence that life is an ever unfolding miracle, that life is beautiful and fleeting.

Our new couples therapist is not a watermelon fetishist. And while admittedly she is a tad zaftig, she is not obese. Not that it matters. Really. She is a heterosexual grand-mother with the slow, careless demeanor of a pothead and the acuity of a television detective. The potheadedness appeals to Faith.

"It's like she takes a big bong hit before she sees us," Faith says without disapproval.

Her flagrant and shrewd critique of our relationship appeals to me. If we're going to have a baby in less than two months, we don't have time for a long, in-depth, feel-good analysis. I want her to get to it—the problem in our relationship—no holds barred. And she does.

The moral of our lesbian story is that Faith and I are very different, like black and white, like network television and HBO, like herbal tea and café au lait. Not that that's news. I mean, it doesn't really require a one-hundred-dollar meeting with a stranger to figure that out. But somehow it makes it all okay, interesting and workable and *okay*. Faith is allowed to go off and become a self-centered bonehead for a day and a half without it meaning she doesn't love or care about me. I'm allowed to demand she try not to go off and become a self-centered bonehead for a day and a half without it meaning I hate her. And so it goes, for five more sessions, the two of us desperately confessing all in order to clear the bases and become as good as we can before our baby arrives.

How many couples go to such lengths before having a baby? Gay people are prohibited from adopting children in so many places and are frowned upon, or worse, as parents in so many more. But how many other parents-to-be race into therapy in the final hour, afraid they won't be at their emotional best in time for the arrival of their new baby? It has nothing to do with anything, really, but occurs to me one morning as I'm reading the paper and just pisses me off, that's all.

Our Japanese Mishpacha

Several tubes of blood and one fasting glucose tolerance test later, it seems I may have gestational diabetes. I was hoping for anemia. That would have been easy. All I would have had to do is take iron pills or continue eating red meat. Instead my high glucose means I have to carbo-load for two days, then fast for twelve hours, drink another sickly sweet drink, and sit for three hungry and tedious hours having my blood drawn, all the while hoping I don't have another one of my sugar-induced spells or a temporary endocrine disorder. Things had been going so smoothly, and while gestational diabetes is hardly a big deal, I just can't help thinking this is all some kind of stupid mistake.

At 8:30 in the morning, one day after our nation's easiest carbo-loading event—Thanksgiving—I arrive at a windowless lab and begin the drinking, peeing, and bloodletting routine that will make up my post-holiday morning. I'm there sitting nervously, waiting to see if the sickly sweet drink will make my pulse shoot up to 120, when four very distinguished Japanese adults enter the waiting area. The group consists of two women in Western clothing and two men, each of whom sports a neatly shaven head and long black ceremonial robe.

Because I'm a bit light-headed and in need both of distraction as well as some—any!—explanation for what is going on with my blood chemistry, I come up with the extraordinarily narcissistic notion that the reason all of this is happening, the reason why recent breakfasts have practically knocked me unconscious, is that I was meant to be here this morning. Not to be excessively ephemeral, but my idea is that the Japanese portion of my child's spiritual ancestry has united me with this group so that they may see how we (me and their baby) are doing. I am but the baby's vessel, after all, her little Jewish guardian, and there are certain eastern goyish spirits out there who are connected to this girl and need to have a look. Is it that outrageous a possibility, the idea that fate wanted to unite me and my unborn child with our Japanese *mishpacha*? It seems at least as feasible as my having gestational diabetes. Besides, everything happens for a reason.

The older of the two Japanese women is in her sixties. It looks like she and the two Buddhist monks (Who other than Buddhist monks would come to my spiritual aid?) are here to have their blood drawn. While the younger of the two women, a very beautiful woman in her fifties, dressed in a salmon-colored leather jacket and black pants, is here as their interpreter.

After leading the older woman into the lab, the beautiful woman returns to wait with the monks. When she takes the seat next to mine I'm neither surprised nor afraid. I'm open and receptive. I'm blasted on sugar, patiently awaiting my spiritual instructions. Discreetly I wipe my upper lip of any trace of a fluorescent orange mustache.

In preparation for my three-hour stay I've made myself quite at home, and suddenly the little world I've created for myself seems degenerate and downright noxious. Slouched over with legs crossed, a book propped against my knee, empty bottles of water at my side, a newspaper, backpack, and balled-up fleece jacket lying on the table to my left, I am positively slovenly in comparison to my daughter's guardian angels. I'm not well-dressed nor well-postured, nor enlightened enough to don a ceremonial robe. I'm pale, hungry, and perhaps temporarily diabetic. While this group eats steamed rice and sea vegetables, I hoard carbohydrates and have myself injected with the store-bought sperm of one of their precious youth. I flash to the last foods I ate before beginning my fast (Cool Whip and leftover turkey), the last gift I had to offer another member of my species (the finger and a sharp toot of the horn on my way to the clinic).

The beautiful woman looks over at me and smiles.

Perhaps she pities me and my freakish Anglo-American ways.

I smile back.

"How far along are you?" the beautiful woman asks.

I can't believe she knows I'm pregnant, what with the nubby, drab olive-green cardigan that's hiding my belly. But then again she was sent to me, and spiritual guidance does not work very well if none of the parties involved have any clue as to what is going on.

"Almost seven months."

The woman doesn't hesitate to mask her surprise. She glances at my stomach, as if upset that I am not eating enough. If she only knew.

"It's hiding," I chuckle, though my immense Jewish wit

suddenly seems crass and vulgar, a far cry from tai chi and green tea.

As the beautiful woman nods, the older woman returns from the lab pulling down her sleeve and looking relieved, thus prompting my friend to rise and escort one of the monks out of the room to the waiting phlebotomist. When she returns, rather than taking her seat, the beautiful woman stands in front of me and gives me a thorough once-over. Like my yoga instructor, without words she motions for me to change my position—uncross your legs.

It doesn't offend me, her command, because I've been waiting for this, their message. I immediately oblige by uncrossing my legs and sitting up straight, forcing my overloaded body into a precise ninety-degree angle. When finally the beautiful woman speaks, I'm perfectly erect, convinced I'm breathing more deeply, feeling more at peace with myself and the world around me, less likely to pass out at any moment.

Still, she furrows her brow and admonishes, "You shouldn't cross your legs when you're pregnant."

"I know," I say, even though I didn't.

"You need to…" She straightens her back as an example, rather than verbally accuse me of poor posture.

"I know." I cinch myself up even straighter.

"And breathe," she says.

I take a big visible breath and tell her I practice yoga. It's embarrassing. "*I* practice *yoga!*" But it slips out just the same.

"Good," she says.

Just then monk number one returns from his blood draw and the whole group gathers its regal, perfect-postured, clean-living selves up from their seats to leave.

That's it? What about the words of wisdom, the secret of the universe? Doesn't monk number two need a blood draw?

Apparently not. My daughter's spiritual kin put on their coats. Oh, well. Trying not to stare at the beautiful woman, I casually return to my book, careful not to reveal its title—*Sex and Real Estate*—while also holding it close enough to my upright face to read the print. Slouching definitely is easier on the eyes.

I sneak a peek whenever possible at the group, letting their lilting Japanese permeate the layers of cotton and epidermis to reach my baby's quarter-Japanese ears. Quarter-Japanese? My own daughter already is so different from her mother. She's her own person with her own ethnicity, her own likes and dislikes, her own fears and desires.

With her hand on the door the beautiful woman calls out to me, "Good luck."

I want to say, "Don't worry, your genes are in good hands." But only a crass Jewish lesbian would insist on pressing this serendipitous point. So, "Thank you," I say, and leave it at that. And then they are gone.

Three days later I learn I don't have gestational diabetes. I smile a serene Buddhist smile as the nurse tells me.

Of course I don't. I could have told you *that*.

Miracles, Maternity Leaves, and a Dear Sweet Talking Baby

I dream the baby is born. She is small like a baby but has hair and talks and is fully dressed.

"How was it for you?" she asks me.

"Labor?"

"Yes."

"It was fine. Really, not bad. How about you?"

"It was good. I'm doing fine."

I'm thinking that while it's such a relief to have a baby who talks already and can communicate her feelings, I will kind of miss the baby stage, having a little baby who cuddles in your arms and gurgles and peeps.

Systemic chemo is working. Crazy and as unlikely as it is, systemic chemo brings my mother's CA-125 down from 7,000 to 3,500. The normal range for this particular marker of ovarian cancer is between zero and thirty-five. When my mother first was diagnosed, her CA-125 was sixty-nine, and we were so freaked out. The surgeon tried to reassure us that sixty-nine was not as bad a number as it could be. It was bad, certainly, but not disastrous. But *Sixty-nine!* we thought. *Cancer!* we thought. How much worse could it be?

My mother isn't really doing better with regard to energy and appetite, but her doctor is thrilled and my mom is thrilled and therefore so are we. The numbers hardly matter, because it's working! It's working! Though I hadn't wanted to think of it in this light, in this immediate and dire light, it's starting to feel like a race to the finish, my mother's life and this pregnancy racing together to beat cancer. But at least the finish line is in sight now. At least we all might win.

"Now stop worrying," my mother tells me. "And give me that grandchild."

For a moment I wonder if she's lying. Maybe the chemo isn't working. Maybe she is convinced once again that denial and secrecy and even lying are better for her daughters than the horrible truth. I consider calling the kind and gentle lesbian oncologist again but decide not to.

"Okay," I tell my mother. After all, we are less than two months away. Less than two months to go! And my mother is doing better—at least her blood is. "But don't rush me," I offhandedly tell my mom. "I've got six more weeks of freedom."

She smiles a sweet, tired smile.

At thirty five and a half weeks' gestation it occurs to me that there are only two things in this life that I know for sure. The first is that one day I will die. The other is that one day I will have this baby. Needless to say, I'm hoping they will not happen on the same day.

It's a powerful existential or Zen or psychotic state to be in, to have absolutely no idea how either of these events will play out and still go about life at work or the supermarket or the vacuum cleaner repair shop. I'm relating normally to people while all the time my mind is circling around the idea, "birth and death, birth and death, birth and death." I'm sure it has to do with the mind-altering combination of my mother's illness and my pregnancy that I am overcome by the idea that everything else in life is up for grabs but for birth and death, birth and death, birth and death.

Faith keeps dreaming about our baby girl. She dreams that she has been born and is calm and gentle and sweet. She dreams of an easy baby, a beautiful baby. She dreams of a four-year-old girl with curly brown hair riding on her shoulders at a park.

I dream that I'm having an affair with a heavyset woman with thick brown hair. She is a bit shy but kind. I reach my arms out to her and pull her into my chest, stroke her hair. I'm not in love, but I like her and it's fun, so what the heck.

At thirty-eight weeks, my mother is going out for dinner with friends and taking on new therapy patients, and I no longer can get out of bed on my own. It's not the weight that's difficult for me to hoist, thirty-five pounds of fluid and baby and placenta and Mocha Chip Balance Bars. It's that my ribs are killing me. Each night when I lie down to sleep, it's as if the baby is trying to pry my ribs apart with a crowbar. Sleeping is impossible. There is no position in which I can hide where our little carpenter with her pointy tools can't find me. Getting out of bed involves tolerating searing pain long enough to get myself in an upright position. Faith extends a hand to help me, but each time I wind up pulling her down with me.

I find myself lying in bed night after sleepless night looking forward to being awakened by a crying baby. At least then I will be able to hoist myself out of bed and tend to her. It's got to be easier being sleepless and not pregnant rather than sleepless *and* pregnant. So what if I'm up half the night, at least I'll be able to turn over. At least I will be able to get out of bed, and tie my own shoes. At least I'll sleep soundly when I do sleep at all.

Nature has a wonderful way of taking care of itself. Imagine if in the weeks before childbirth an expectant mother was sleeping better and longer than ever before. Imagine if being eight or nine months pregnant did not involve swollen feet and legs, constipation, and uncontrollable gas. The arrival of a small, vulnerable, nocturnal newborn would be even more like jumping into a pool of ice than it already is.

With just days left in this incredible journey called pregnancy my sister and Kevin and Jonah arrive for a visit. Despite a softening cervix and the loss of something that rather closely resembled a mucus plug, it doesn't look like I'll be giving birth during their visit. (Hours, days, or sometimes weeks before a woman delivers, she often passes her mucus plug. A half-finger in length, the mucus plug looks a lot like, well, mucus, and once it comes out it seems almost impossible that your baby will not in the next few seconds just shoot out of you like greased lightning. To a first-time mother, the passing of the mucus plug is much like a cork coming out of a wine bottle—it's like, Oh, boy, stop the presses, here it comes. Only it doesn't work like that because as it turns out there's a hell of a lot more than a wad of mucus between in and out, between being a fetus and becoming a newborn, like, for instance, *labor*. When asked by a woman what she should do if it seems the baby is coming out on the way to the hospital or in the house, God forbid, our childbirth instructor answered, "Thank your lucky stars," i.e., it takes more than the passing of a mucus plug to pop a baby. But that is not how it feels to me, sitting on my mother's sofa holding my nephew on my mucus plug–less lap. It feels like our daughter, Jonah's first cousin, might be slipping out of me any minute now, and then there would be not one but two babies here in my mother's apartment and that seems totally crazy and unreasonable.)

Carrie and Kevin are charting everything. They keep a detailed record of Jonah's poops and pees, the number of ounces of milk he has drank, the start and stop times of each nap. It seems like so much work. Faith and I didn't know about the chart part of parenting. We look on with trepidation.

"What if we're too tired to remember when she's pooped?" I ask, terrified I may not have what it takes to be a good record-keeping mother.

"Oh, that will happen. It's okay if you forget a few."

I tell Faith we need graph paper. It's just one more item for the long list of things we MUST get before our daughter is born: Desitin, alcohol- and fragrance-free wipes, petroleum jelly, baby shampoo, and now graph paper. And where will we keep our chart? Should we have a hook on the wall near the changing table from which we might hang a clipboard, or should we invest in a dry-erase board?

Meanwhile Jonah smiles and gurgles and farts out loud. He is a handsome little guy, with a head full of blond hair and a big gummy smile, and he seems entirely unconcerned with all of the data he generates. He just eats and sleeps and poops and eats and sleeps and pees and poops all day long, while Kevin and Carrie run ragged trying to keep him from crying, making sure he sleeps enough, and preparing him bottle after bottle.

"He's eating again? You fed him two hours ago." I'm just pregnant. What do I know.

"Yeah," says my sister. Duh.

I can't get over the rapid repetition of the eat-pee-poop cycle. It's like all day long you just feed and change a baby. Feed and change. Feed and change. Feed and change. That this will soon be our life I cannot imagine. That this baby is my nephew and that there is a very similar model growing inside me I cannot imagine. When I have to assist my sister in a diaper change on the floor of a handicapped stall in a public rest room while Jonah screams bloody murder, I cannot imagine any similar scenario coming soon to a theater near me. That five months from now I will have what I think is the incredibly creative idea of keeping a list of all of the places I have changed my daughter's diaper and/or nursed her is an image of the future that does not even come close to crossing my very pregnant mind.

While my sister and Kevin are in sleepless hell, my mother is in absolute heaven. She has her grandson under her roof, a granddaughter on the way, and continued good reports from her oncologist. It all makes up for the fact that she is still losing weight and needing to nap every afternoon.

Carrie confesses to having had the same paranoid thought as me, that our mother is lying to us about her diminishing CA-125 and normal liver function tests. We ask her friends and they tell us that our mother has been reporting to them the very same information. And so we file away our dread and continue to take turns lifting Jonah up toward the ceiling higher and higher until he squeals with delight. I make dinner and Carrie and Faith clean up. My mother holds Jonah on her lap—she is not able to stand and hold him at the same time—and rocks him to sleep. Our daughter tumbles inside me, anxious to get out and join the party.

At forty weeks I am two centimeters dilated, unquestionably free of my mucus plug, big, round, and very, very pregnant. Still, I continue to drag myself in to work each day. Despite constant fatigue, swollen ankles, and an inability to wipe my own arse, I'll be damned if I use up any portion of my precious vacation and sick time lying around the house. While the hospital at which I am employed is legally obligated to provide women employees with a three-month maternity leave, they are not, unfortunately, required to pay employees for that time. Instead, women must use whatever vacation and sick hours they have accumulated to support themselves during their absence. It's a wonderful system. At the end of three months you must return to work at whatever hourly commitment you held prior to delivering your baby or risk losing your job.

I have no intention of returning to work full-time when our daughter is three months old. While it's difficult to imagine *ever* returning to work full-time, I decide that when our daughter is six months old I will miraculously be able to do just that. I convey this to my boss, a man with whom I have been working for the last fifteen years, and in a moment of compassion he suggests this unbelievable arrangement: from birth to three months I will not work at all, from three to six months I will come in four to eight hours per week.

From the vantage point of a pregnant forty-year-old who has worked full-time almost every day of her life since graduating from college, this seems amazing, ridiculous even. What's four to eight hours a week? It's like the blink of an eye, a piss in the ocean, a baby's afternoon nap. Four to eight hours per week? Ha! I'm so psyched. Piece of cake! After six months I assume our daughter will be such a secure and independent baby, and Faith and I will be such secure and independent mommies, that entering her into day care will be a breeze. I mean, it's not like she'll be only three months old! I'm made in the shade. Lucky as the devil. Who needs more than six months at home? Besides, it's not like I'm a *baby person* or anything.

That I'll have no health insurance and very little income during months three to six is only minimally distressing. Around the time we decided to venture forth and have a family, Faith and I started to squirrel away as much money as we could to support an extended maternity leave. What else is money for if not to allow a mother to be with her newborn child? With Faith still working, we'll have enough to cover insurance and living expenses through month six.

I have no idea how revolutionary an act this is until other mothers ask, "How long will you be at home?" Women with far more money and resources than Faith and I ever may have (i.e., their husband's health insurance policy, and careers other than those in the arts or academia) respond to my answer of "six months" with "Oh, you're so *lucky*. I went back after three." That my maternity leave has nothing whatsoever to do with luck and everything to do with priorities does not dawn on me until months into it, when I'm a mother and our daughter at six months feels just as small and vulnerable as she did at three, and suddenly everything about being a parent reveals itself to be an endless series of choices, surprises, and vastly creative solutions.

But that's much later, really. That's after giving birth to our daughter, after quitting my job entirely, after watching my mother die. If there are two events that immediately throw one's life into perspective, they are becoming a parent and losing your mother. So much for therapy. So much for years of pondering my own navel. That I both gained a daughter and lost a mother during the same three months of my life, well, maybe that was luck. Bad luck and good luck wrapped in and around each other in some crazy earth-shattering double helix. Birth and death, birth and death, birth and death.

By then I was managing. By then, three sleepless months after our daughter was born, I was a person who could manage. Suddenly and instantaneously I was a woman who could mother, a daughter who somehow could survive without her own mother.

But that's later, really. So much later. For now I'm pregnant. Very, very pregnant. Pregnant and *waiting*. *Tick tick tick.*

Epilogue

Betsy Grace was born at 6:27 on a crisp morning in February. Betsy waited a full four days after her due date to finish packing and let her mommies know that she was ready at last to try living on the outside. That she arrived via a moderately urgent C-section did not diminish the joy or exquisite splendor her family experienced upon meeting her for the first time. In fact, the extra three days in the hospital the C-section afforded her mommies was much appreciated—not to mention her birth mother's intact perineum.

Because there are miracles and magic in the world, Betsy's grandmother was able to hold her in her arms, to know her, and to love her during the first months of Betsy's life and the last few months of her own. And because there are babies and children in the world, even after loss and pain miracles and magic continue.

ACKNOWLEDGMENTS

I cannot thank enough Ann Collette of the Helen Rees Agency, Angela Brown of Alyson Publications, and Jon Marcus of *Boston Magazine* for their enthusiasm and support of this manuscript in all of its various stages and incarnations.

A great big thank you to Gail Leondar-Wright, whose energy and wisdom late in the game helped rouse me from the depths of sleep-deprivation and mourning to finish this project.

I am forever indebted to my most original traveling companions, Sandra Aizley, Robert Aizley, and Carolyn Aizley Neustadt, for their continued love and patience.

Support: to buoy up, to maintain, to sustain, to keep up. And for that I thank always Victoria Lane, Sohaila Abdulali, and Geoff Alperin.

It's impossible to imagine the creation of this book, the creation of our daughter, or the creation of many miraculous tomorrows without my partner, Faith Soloway.

Finally, thank you to all of the unknown men* who—for fun or for money, on a dare or on principle—have made so many women's dreams of motherhood come true. We are taking good care of the children.

*All donor I.D. numbers as well as any identifying traits and characteristics have been changed, masked, and distorted beyond recognition to protect the privacy of all involved.

ABOUT THE AUTHOR

FAITH SOLOWAY

Harlyn Aizley's writing has appeared in numerous literary journals and magazines, including *96 Inc., Berkeley Fiction Review, Boston Magazine, Mangrove,* and *The South Carolina Review,* as well as in the anthologies *Beginnings, Love Shook My Heart,* and *Scream When You Burn.* Harlyn is a graduate of Brandeis University and the Harvard Graduate School of Education. A native of New Jersey, she currently lives in the Boston area.